CYCLES OF TIME

Professor Sir Roger Penrose is Emeritus Rouse Ball Professor of Mathematics at the University of Oxford. He has received a number of prizes and awards, including the 1988 Wolf Prize for physics which he shared with Stephen Hawking for their joint contribution to our understanding of the universe.

ALSO BY ROGER PENROSE

The Emperor's New Mind: Concerning Computers, Minds, and the Laws of Physics

Shadows of the Mind: A Search for the Missing Science of Consciousness

Road to Reality: A Complete Guide to the Laws of the Universe

ROGER PENROSE

Cycles of Time

An Extraordinary New View of the Universe

VINTAGE BOOKS
London

Published by Vintage 2011

4 6 8 10 9 7 5 3

First published in Great Britain in 2010 by
The Bodley Head

Vintage
Random House, 20 Vauxhall Bridge Road,
London SW1V 2SA

www.vintage-books.co.uk

Addresses for companies within The Random House Group Limited can be found at: www.randomhouse.co.uk/offices.htm

The Random House Group Limited Reg. No. 954009

A CIP catalogue record for this book is available from the British Library

ISBN 9780099505945

The Random House Group Limited supports The Forest Stewardship Council (FSC®), the leading international forest certification organisation. Our books carrying the FSC label are printed on FSC® certified paper. FSC is the only forest certification scheme endorsed by the leading environmental organisations, including Greenpeace. Our paper procurement policy can be found at www.randomhouse.co.uk/environment

Typeset by Palimpsest Book Production Ltd,
Falkirk, Stirlingshire
Printed and bound by CPI Group (UK) Ltd, Croydon, CR0 4YY

Roger Penrose's groundbreaking and bestselling *The Road to Reality* (2005) provided a comprehensive yet readable guide to our understanding of the laws that are currently believed to govern our universe. In *Cycles of Time*, he moves far beyond this to develop a completely new perspective on cosmology, providing a quite unexpected answer to the often-asked question 'What came before the Big Bang?'

The two key ideas underlying this novel proposal are a penetrating analysis of the Second Law of thermodynamics—according to which the 'randomness' of our world is continually increasing—and a thorough examination of the *light-cone* geometry of space-time. Penrose is able to combine these two central themes to show how the expected ultimate fate of our accelerating, expanding universe can actually be reinterpreted as the 'big bang' of a new one.

On the way, many other basic ingredients are presented, and their roles discussed in detail, though without any complex mathematical formulae (these all being banished to the appendices). Various standard and non-standard cosmological models are presented, as is the fundamental and ubiquitous role of the cosmic microwave background. Also crucial to the discussion are the huge black holes lying in galactic centres, and their eventual disappearance via the mysterious process of Hawking evaporation.

Contents

Preface

ONE of the deepest mysteries of our universe is the puzzle of whence it came.

When I entered Cambridge University as a mathematics graduate student, in the early 1950s, a fascinating cosmological theory was in the ascendant, known as the steady-state model. According to this scheme, the universe had no beginning, and it remained more-or-less the same, overall, for all time. The steady-state universe was able to achieve this, despite its expansion, because the continual depletion of material arising from the universe's expansion is taken to be compensated by the continual creation of new material, in the form of an extremely diffuse hydrogen gas. My friend and mentor at Cambridge, the cosmologist Dennis Sciama, from whom I learnt the thrill of so much new physics, was at that time a strong proponent of steady-state cosmology, and he impressed upon me the beauty and power of that remarkable scheme of things.

Yet this theory has not stood the test of time. About 10 years after I had first entered Cambridge, and had become well acquainted with the theory, Arno Penzias and Robert Wilson discovered, to their own surprise, an all-pervading electromagnetic radiation, coming in from all directions, now referred to as the *cosmic microwave background* or CMB. This was soon identified, by Robert Dicke, as a predicted implication of the 'flash' of a *Big-Bang* origin to the universe, now presumed to have taken place some 14 thousand million years ago—an event that had been first seriously envisaged by Monsignor Georges Lemaître in 1927, as an implication of his work on Einstein's 1915 equations of general relativity and early observational indications of an expansion of the universe. With great courage and scientific honesty (when the CMB data became better established), Dennis Sciama publicly repudiated his earlier views and strongly supported the idea of the Big Bang origin to the universe from then on.

Since that time, cosmology has matured from a speculative pursuit into an exact science, and intense analysis of the CMB—coming from highly detailed data, generated by numerous superb experiments—has formed a major part of this revolution. However, many mysteries remain, and much speculation continues to be part of this endeavour. In this book, I provide descriptions not only of the main models of classical relativistic cosmology but also of various developments and puzzling issues that have arisen since then. Most particularly, there is a profound oddness underlying the Second Law of thermodynamics and the very nature of the Big Bang. In relation to this, I am putting forward a body of speculation of my own, which brings together many strands of different aspects of the universe we know.

My own unorthodox approach dates from the summer of 2005, though much of the detail is more recent. This account goes seriously into some of the geometry, but I have refrained from including, in the main body of the text, anything serious in the way of equations or other technicalities, all these being banished to the Appendices. The experts, only, are referred to those parts of the book. The scheme that I am now arguing for here is indeed unorthodox, yet it is based on geometrical and physical ideas which are very soundly based. Although something entirely different, this proposal turns out to have strong echoes of the old steady-state model!

I wonder what Dennis Sciama would have made of it.

Acknowledgements

I am very grateful to many friends and colleagues for their important inputs, and for sharing their thoughts with me relating to the cosmological scheme that I am putting forward here. Most importantly, detailed discussions with Paul Tod, concerning the formulation of his proposal for a conformal-extension version of the Weyl curvature hypothesis have been crucially influential, and many aspects of his analysis have proved vital to the detailed development of the equations of conformal cyclic cosmology, as I am putting them forward here. At the other end of things, Helmut Friedrich's powerful analysis of conformal infinity, particularly his work on the case where there is a positive cosmological constant, has lent a strong support to the mathematical viability of this scheme. Another who has contributed an important input over a great many years is Wolfgang Rindler, particularly for his seminal understanding of cosmological horizons, but also for his long collaboration with me on the 2-spinor formalism, and also for discussions on the role of inflationary cosmology.

Significant input has come from Florence Tsou (Sheung Tsun) and Hong-Mo Chan for sharing with me their ideas concerning the nature of mass in particle physics, and James Bjorken also provided a crucial insight in relation to this. Among the many others who have importantly influenced me are David Spergel, Amir Hajian, James Peebles, Mike Eastwood, Ed Speigel, Abhay Ashtekar, Neil Turok, Pedro Ferreira, Vahe

Gurzadyan, Lee Smolin, Paul Steinhardt, Andrew Hodges, Lionel Mason, and Ted Newman. Richard Lawrence's heroic editorial support has been invaluable, as has the vital input from Thomas Lawrence in providing much missing information, particularly concerning Part 1. Thanks are due to Paul Nash for indexing. I am grateful also to the translators Daniele Didero and Katharina Neusser for pointing out various misprints and minor errors in the earlier printings.

For her deep support, love and understanding, under frequently difficult circumstances, I am profoundly indebted to my wife Vanessa, whom I thank also for immediately providing some needed graphs at short notice, but more particularly for guiding me through some of the continual frustrations of modern electronic technology that would otherwise have completely defeated me with regard to the diagrams. Finally, our 10-year-old son Max is to be thanked not only for his continual encouragement and good cheer, but also for playing his own part in helping me through this bewildering technology.

I am grateful to the M.C. Escher Company, Holland, for permission to reprint the images used in Fig. 2.3. Thanks go to the Institute of Theoretical Physics, University of Heidelberg for Fig. 2.6. I give thanks, in addition, to NSF for support under PHY00-90091.

Prologue

WITH his eyelids half closed, as the rain pelted down on him and the spray from the river stung his eyes, Tom peered into the swirling torrents as the water rushed down the mountainside. 'Wow', he said to his Aunt Priscilla, an astrophysics professor from the University of Cambridge, who had taken him to this wonderful old mill, preserved in excellent working order, 'is it always like this? No wonder all that old machinery can be kept buzzing around at such great speed.'

'I don't think it's always this energetic', said Priscilla, standing next to him behind the railing at the side of the river, and raising her voice somewhat, so as to be heard over the noise of the rushing water. 'The water's much more violent than usual, today, because of all this wet weather. You can see down there that a good portion of the water has had to be diverted away from the mill. Usually they would not do this, because they would have to make the most of a much more sedate flow. But now there's far more energy in the flow than is needed for the mill.'

Tom stared for some minutes into the wildly tumbling water and admired the patterns it made as it was flung into the air in sprays and convoluted surfaces. 'I can see there's a lot of power in that water, and I know that a couple of centuries ago the people were clever enough to see how all this energy could be used to drive these machines—doing the work of many human beings and making all that great woollen cloth. But where did the energy come from that got all that water high up on the mountain in the first place?'

'The heat of the Sun caused the water in the oceans to evaporate and rise up into the air, so it would eventually come back down again in all this rain. So a good proportion of the rain would be deposited up high into the mountains', replied Priscilla. 'It's really the energy from the Sun that is being harnessed to run the mill.'

Tom felt a little puzzled by this. He was often puzzled by the things that Priscilla told him, and was by nature often quite sceptical. He could not really see how just heat could lift water up into the air. And if there was all that heat around, why did he feel so cold now? 'It was rather hot yesterday', he grudgingly agreed. Though, still uneasy, he commented, 'but I didn't feel the Sun trying to lift me up into the air then, any more than I do now.'

Aunt Priscilla laughed. 'No. it's not really like that. It's the tiny little molecules in the water in the oceans that the Sun's heat causes to be more energetic. So these molecules then rush randomly around faster than they would otherwise, and a few of these "hot" molecules will move so fast that they break loose from the surface of the water and are flung into the air. And although there are only a relatively few molecules flung out at one time, the oceans are so vast that there would really be a lot of water flung up into the air altogether. These molecules go to make the clouds and eventually the water molecules fall down again as rain, a lot of which falls high in the mountains.'

Tom was still rather troubled, but at least the rain had now tapered off somewhat. 'But this rain doesn't feel at all hot to me.'

'Think of the Sun's heat energy first getting converted into the energy of rapid random motion of the water molecules. Then think of this rapid motion resulting in a small proportion of the molecules going so fast that they are flung high in the air in the form of water vapour. The energy of these molecules gets converted into what's called gravitational potential energy. Think of throwing a ball up into the air. The more energetically you throw it the higher it goes. But when it reaches its maximum height, it stops moving upwards. At that point its energy of motion has all been converted into this gravitational potential energy in its height above the ground. It's the same with the water molecules. Their energy of motion—the energy that they got from the Sun's heat—is converted into this gravitational potential energy, now at the top of the mountain, and when it runs down, this is converted back again into the energy in its motion, which is used to run the mill.'

'So the water isn't hot at all when it's up there?' asked Tom.

'Exactly, my dear. By the time that these molecules get very high in the sky, they slow down and often actually get frozen into tiny ice crystals—that's what most clouds are made of—so the energy goes into their height above the ground rather than into their heat motion. Accordingly, the rain won't be hot at all up there, and it's still quite cold even when it finally works its way down again, slowed down by the resistance of the air.'

'That's amazing!'

'Yes, indeed', and encouraged by the boy's interest, Aunt Priscilla eagerly took advantage of the opportunity to say more. 'You know, it's a curious fact that even in the *cold* water in this river there is still much more *heat* energy in the motion of the individual molecules running around randomly at great speed than there is in the swirling currents of water rushing down the mountainside!'

'Goodness. I'm supposed to believe that, am I?'

Tom thought for a few minutes, somewhat confused at first, but then rather attracted by what Priscilla said, remarked excitedly: 'Now you've given me a great idea! Why don't we build a special kind of mill that just directly uses all that energy of the motion of water molecules in some ordinary lake? It could use lots of tiny little windmill things, maybe like those things that spin in the wind, with little cups on the ends so that they twirl round in the wind no matter which direction the wind is coming from. Only they'd be very tiny and in the water, so that the speed of the water molecules would spin them around, and you could use these to convert the energy in the motion in the water molecules to drive all sorts of machinery.'

'What a wonderful idea, Tom darling, only unfortunately it wouldn't work! That's because of a fundamental physical principle known as the Second Law of thermodynamics, which more or less says that things just get more and more disorganized as time goes on. More to the point, it tells you that you can't get useful energy out of the *random* motions of a hot—or cold—body, just like that. I'm afraid what you're suggesting is what they call a "Maxwell's demon".'

'Don't *you* start doing that! You know that Grandpa always used to call me a "little demon" whenever I had a good idea, and I didn't like it. And, that Second Law thing's not a very nice kind of law', Tom complained grumpily. Then his natural scepticism returned: 'And I'm not sure I can really believe in it anyway.' Then he continued 'I think laws like that just need clever ideas to get around them. In any case, I thought you said that it's the heat of the Sun that's responsible for heating the oceans and that it's that *random* energy of motion that flings it to the top of the mountain, and that's what's running the mill.'

'Yes, you're right. So the Second Law tells us that actually the heat of the Sun all by itself wouldn't work. In order to work, we *also* need the colder upper atmosphere, so that the water vapour can condense up above the mountain. In fact, the Earth as a whole doesn't get energy from the Sun overall.'

Tom looked at his aunt with a quizzical expression. 'What does the cold upper atmosphere have to do with it? Doesn't "cold" mean not so much energy as "hot"? How does a bit of "not-so-much energy" help? I don't get what you are saying at all. Anyway, I think you are contradicting yourself', said Tom, gaining confidence in himself. 'First you tell me that the Sun's energy runs the mill, and now you tell me that the Sun doesn't give energy to the Earth after all!'

'Well, it doesn't. If it did, then the Earth would just keep on getting hotter and hotter as it gained energy. The energy that the Earth gets from the Sun in the daytime has all to go back into space eventually, which it does because of the cold night sky—except, I suppose, that with global warming, a little part of it does get held back by the Earth. It's because the Sun is a very hot spot in an otherwise cold dark sky . . .'

Tom began to lose the thread of what she was saying and his mind began to wander. But he heard her say, '. . . so it's the manifest *organization* in the Sun's energy that enables us to keep the Second Law at bay.'

Tom looked at Aunt Priscilla, almost totally bemused. 'I don't think I really understand all that,' he said, 'and I don't see why I need to believe that "Second Law" thing in any case. Anyway, where does all that organization in the Sun come from? Your Second Law should be telling us that the Sun's getting more disorganized as time goes on, so it would have to have been enormously organized when it was first formed, since all the time it's sending out organization. Your "Second Law" thing tells us that its organization keeps getting lost.'

'It has to do with the Sun being such a hot spot in a dark sky. This extreme temperature imbalance provided the needed organization.'

Tom stared at Aunt Priscilla, with little comprehension, and now not really properly believing anything she was telling him. 'You tell me that counts as organization; well, I don't see why it should. All right, let's pretend it somehow does—but then you still haven't told me where that funny kind of organization comes from.'

'From the fact that the gas that the Sun condensed from was previously spread uniformly, so that gravity could cause it to form clumps which condensed gravitationally into stars. A very long time ago, the Sun did just this; it condensed from this initially spread-out gas, getting hotter and hotter in the process.'

'You'll keep telling me one thing after another, going way back in time, but where does this thing you call "organization", whatever it is, *originally* come from?'

'Ultimately it comes from the Big Bang, which was what started the whole universe off with an utterly stupendous explosion.'

'A thing like a big walloping explosion doesn't sound like something organized. I don't get it at all.'

'You aren't the only one! You're in good company not to get it. Nobody *really* gets it. It's one of the biggest puzzles of cosmology where the organization comes from, and in what way the Big Bang really represents organization in any case.'

'Maybe there was something *more* organized *before* the Big Bang? That might do it.'

'People have actually tried suggesting things like that for some while. There are theories in which our presently expanding universe had a previous collapsing phase which "bounced" to become our Big Bang. And there are other theories where little bits of a previous phase of the universe collapsed into things we call black holes, and these bits "bounced", to become the seeds of lots and lots of new expanding universes, and there are others where new universes sprang out of things called "false vacuums". . .'

'That all sounds pretty crazy to me,' Tom said.

'And, oh yes, there's another theory that I heard about recently . . .'

Cycles of Time

An Extraordinary New View of the Universe

Part 1

The Second Law and its underlying mystery

1.1 The relentless march of randomness

THE Second Law of thermodynamics—what law is this? What is its central role in physical behaviour? And in what way does it present us with a genuinely deep mystery? In the later sections of this book, we shall try to understand the puzzling nature of this mystery and why we may be driven to extraordinary lengths in order to resolve it. This will lead us into unexplored areas of cosmology, and to issues which I believe may be resolved only by a very radical new perspective on the history of our universe. But these are matters that will be our concern later. For the moment let us restrict our attention to the task of coming to terms with what is involved in this ubiquitous law.

Usually when we think of a 'law of physics' we think of some assertion of equality between two different things. Newton's second law of motion, for example, equates the rate of change of momentum of a particle (momentum being mass times velocity) with the total force acting upon it. As another example, the law of conservation of energy asserts that the total energy of an isolated system at one time is equal to its total energy at any other time. Likewise, the law of conservation of electric charge, of momentum, and of angular momentum, each asserts a corresponding equality for the total electric charge, for the total momentum, and for total angular momentum. Einstein's famous law $E=mc^2$ asserts that the energy of a system is always equal to its mass multiplied by the square of the speed of light.

As yet another example, Newton's third law asserts that the force exerted by a body A on a body B, at any one time, is always equal and opposite to the force acting *on* A due to B. And so it is for many of the other laws of physics.

These are all *equalities*—and this applies also to what is called the *First* Law of thermodynamics, which is really just the law of conservation of energy again, but now in a thermodynamic context. We say 'thermodynamic' because the energy of the *thermal motions* is now being taken into account, i.e. of the random motions of individual constituent particles. This energy is the *heat* energy of a system, and we define the system's *temperature* to be this energy per degree of freedom (as we shall be considering again later). For example, when the friction of air resistance slows down a projectile, this does not violate the full conservation law of energy (i.e. the First Law of thermodynamics)—despite the loss of kinetic energy, due to the projectile's slowing—because the air molecules, and those in the projectile, become slightly more energetic in their random motions, from *heating* due to the friction.

However, the *Second* Law of thermodynamics is not an equality, but an *inequality*, asserting merely that a certain quantity referred to as the *entropy* of an isolated system—which is a measure of the system's disorder, or 'randomness'—is *greater* (or at least not smaller) at later times than it was at earlier times. Going along with this apparent weakness of statement, we shall find that there is also certain vagueness or subjectivity about the very definition of the entropy of a general system. Moreover, in most formulations, we are led to conclude that there are occasional or exceptional moments at which the entropy must be regarded as actually (though temporarily) *reducing* with time (in a fluctuation) despite the general trend being that the entropy increases.

Yet, set against this seeming imprecision inherent in the Second Law (as I shall henceforth abbreviate it), this law has a universality that goes far beyond any particular system of dynamical rules that one might be concerned with. It applies equally well, for example, to relativity theory as it does to Newtonian theory, and also to the continuous fields of Maxwell's theory of electromagnetism (that we shall be coming to briefly in §2.6, §3.1 and §3.2, and rather more explicitly in Appendix A1) just as well as it does to theories involving only discrete particles. It applies also to hypothetical dynamical theories that we have no good reason to believe have relevance to the actual universe that we inhabit, although it is most pertinent when applied to realistic dynamical schemes, such as Newtonian mechanics, which have a *deterministic* evolution and are *reversible in time*, so that for any allowed evolution into the future, reversing the time direction gives us another equally allowable evolution according to the dynamical scheme.

To put things in familiar terms, if we have a moving-picture film depicting some action that is in accordance with dynamical laws—such as Newton's—that are reversible in time, then the situation depicted when the film is run in reverse will also be in accordance with these dynamical laws. The reader might well be puzzled by this, for whereas a film depicting an egg rolling off a table, falling to the ground, and smashing would represent an allowable dynamical process, the time-reversed film—depicting the smashed egg, originally as a mess on the floor, miraculously assembling itself from the broken pieces of shell, with the yolk and albumen separately joining up to become surrounded by the self-assembling shell, and then jumping up on to the table—is not an occurrence that we expect ever to see in an actual physical process (Fig. 1.1). Yet the full Newtonian dynamics of each individual particle, with its accelerated response (in accordance with Newton's second law) to all forces acting upon it, and the elastic reactions involved in any collision between constituent particles, is completely reversible in time. This also would be the case for the refined behaviour of relativistic and quantum-mechanical particles, according to the standard procedures of modern physics—although there are some subtleties arising from the black-hole physics of general relativity, and also with regard to quantum mechanics, that I do not wish to get embroiled in just yet. Some of these subtleties will actually be crucially important for us later, and will be considered particularly in §3.4. But for the moment, an entirely Newtonian picture of things will suffice.

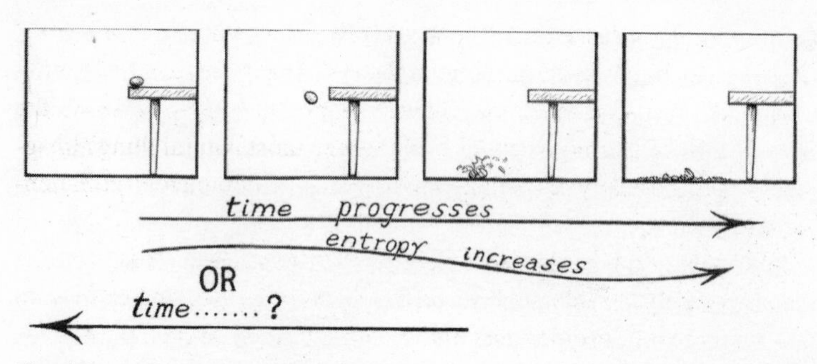

Fig. 1.1 An egg rolling off a table, falling to the ground and smashing according to time-reversible dynamical laws.

We have to accustom ourselves to the fact that the situations that are depicted by *both* directions of film-running are consistent with Newtonian dynamics, but the one showing the self-assembling egg depicts an occurrence that is inconsistent with the Second Law, and would be such an enormously improbable sequence of events that we can simply reject it as a realistic possibility. What the Second Law indeed states, roughly speaking, is that things are getting more 'random' all the time. So if we set up a particular situation, and then let the dynamics evolve it into the future, the system will evolve into a more random-looking state as time progresses. Strictly, we should not say that it *will* evolve into a more random-looking state but that, in accordance with what has been said above, it is (something like) *overwhelmingly likely* to evolve into such a more random state. In practice, we must expect that, according to the Second Law, things are indeed getting progressively more and more random with time, but that this represents merely an overwhelming probability, not quite an absolute certainty.

Nevertheless we can assert, with a considerable amount of confidence, that what we shall experience will be an entropy increase—in other words an increase in randomness. Stated that way, the Second Law sounds perhaps like a council of despair, for it tells us that things are just getting more and more disorganized as time progresses. This does not sound like any kind of a mystery, however, as the title of Part 1 seems to be suggesting that it should. It's just an obvious feature of the way things would behave if left entirely to themselves. The Second Law appears to be just expressing an inevitable and perhaps depressing feature of everyday existence. Indeed, from this point of view, the Second Law of thermodynamics is one of the most natural things imaginable, and certainly something that reflects a completely commonplace experience.

Some might worry that the emergence of life on this Earth, with its seemingly unbelievable sophistication, represents a contradiction with this increase of disorder that the Second Law demands. I shall be explaining later (see §2.2) why there is in fact no contradiction. Biology is, as far as we know, entirely consistent with the overall entropy increase that the Second Law demands. The mystery referred to in the title of Part 1 is a mystery of *physics* of an entirely different order of scale. Although it has some definite relation to that mysterious and puzzling organization that we are continually being presented with through biology, we have good reason to expect that the latter presents no paradox with regard to the Second Law.

One thing should be made clear, however, with regard to the Second Law's physical status: it represents a separate principle that must be *adjoined* to the dynamical laws (e.g. to Newton's laws), and is not to be regarded as a *deduction* from them. The actual *definition* of the entropy of a system *at any one moment* is, however, symmetrical with regard to the direction of time (so we get the same entropy definition, for our filmed falling egg, at any one moment, irrespective of the direction in which the film is shown), and if the dynamical laws are also symmetrical in time (as is indeed the case with Newtonian dynamics), the entropy of a system being not always constant in time (as is clearly so with the smashing egg), then the Second Law cannot be a deduction from these dynamical laws. For if the entropy is increasing in a particular situation (e.g. egg smashing), this being in accordance with the Second Law, then the entropy must be *decreasing* in the reversed situation (egg miraculously assembling), which is in gross violation of the Second Law. Since both processes are nevertheless consistent with the (Newtonian) dynamics, we conclude that the Second Law cannot simply be a *consequence* of the dynamical laws.

1.2 Entropy, as state counting

But how does the physicist's notion of 'entropy', as it appears in the Second Law, actually *quantify* this 'randomness', so that the self-assembling egg can indeed be seen to be overwhelmingly improbable, and thereby rejected as a serious possibility? In order to be a bit more explicit about what the entropy concept actually is, so that we can make a better description of what the Second Law actually asserts, let us consider a physically rather simpler example than the breaking egg. The Second Law tells us, for example, that if we pour some red paint into a pot and then some blue paint into the same pot and give the mixture a good stir, then after a short period of such stirring the different regions of red and of blue will lose their individuality, and ultimately the entire contents of the pot will appear to have the colour of a uniform purple. It seems that no amount of further stirring will convert the purple colour back to the original separated regions of red and blue, despite the time-reversibility of the submicroscopic physical processes underlying the mixing. Indeed, the purple colour should eventually come about spontaneously, even without the stirring, especially if we were to warm the paint up a little. But with stirring, the purple state is reached much more quickly. In terms of entropy, we find that the original state, in which there are distinctly separated regions of red and blue paint, will have a relatively low entropy, but that the pot of entirely purple paint that we end up with will have a considerably larger entropy. Indeed, the whole stirring procedure provides us with a situation that is not only consistent with the Second Law, but which begins to give us a feeling of what the Second Law is all about.

Let us try to be more precise about the entropy concept, so that we can be more explicit about what is happening here. What actually *is* the entropy of a system? Basically, the notion is a fairly elementary one, although involving some distinctly subtle insights, due mainly to the great Austrian physicist Ludwig Boltzmann, and it has to do just with *counting* the different possibilities. To make things simple, let us idealize our pot of paint example so that there is just a (very large) finite number of different possibilities for the locations of each molecule of red paint or of blue paint. Let us think of these molecules as red balls or blue balls, these being allowed to occupy only discrete positions, centred within N^3 cubical compartments, where we are thinking of our paint pot as an enormously subdivided $N \times N \times N$ cubical crate composed of these compartments (see Fig. 1.2), where I am assuming that every compartment is occupied by exactly one ball, either red or blue (represented as white and black, respectively, in the figure).

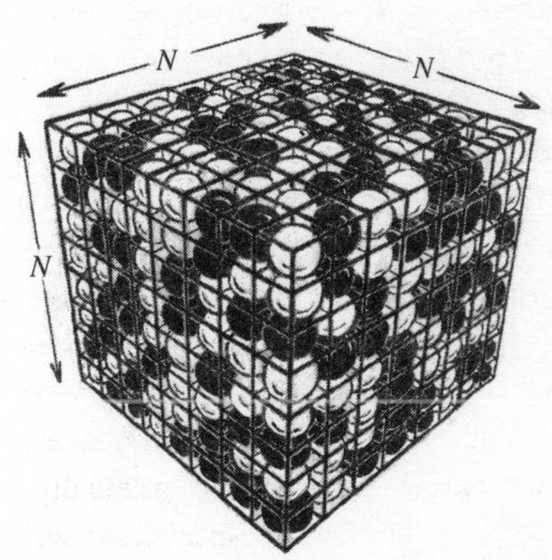

Fig. 1.2 $N \times N \times N$ cubical crate, each compartment containing a red or blue ball.

To judge the colour of the paint at some place in the pot, we make some sort of average of the relative density of red balls to blue balls in the neighbourhood of the location under consideration. Let us do this by containing that location within a cubical box that is much smaller than the entire crate, yet very large as compared with the individual cubical compartments just considered. I shall suppose that this box contains a large number of the compartments just considered, and belongs to a cubical array of such boxes, filling the whole crate in a way that is less refined than that of the original compartments (Fig. 1.3). Let us suppose that each box has a side length that is n times as great as that of the original compartments, so that there are $n \times n \times n = n^3$ compartments in each box. Here n, though still very large, is to be taken to be far smaller than N:

$$N \gg n \gg 1.$$

To keep things neat, I suppose that N is an exact multiple of n, so that

$$N = kn$$

where k is a whole number, giving the number of boxes that span the crate along each side. There will now be $k \times k \times k = k^3$ of these intermediate-sized boxes in the entire crate.

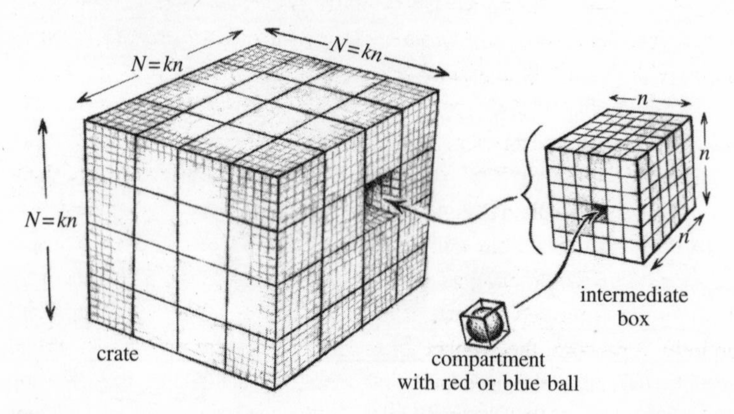

Fig. 1.3 The compartments are grouped together into k^3 boxes, each of size $n \times n \times n$.

The idea will be to use these intermediate boxes to provide us with a measure of the 'colour' that we see at the location of that box, where the balls themselves are considered to be too small to be seen individually. There will be an average colour, or *hue* that can be assigned to each box, given by 'averaging' the colours of the red and blue balls within that box. Thus, if r is the number of red balls in the box under consideration, and b the number of blue balls in it (so $r+b=n^3$), then the hue at that location is taken to be defined by the ratio of r to b. Accordingly, we consider that we get a redder hue if r/b is larger than 1 and a bluer hue if r/b is smaller than 1.

Let us suppose that the mixture *looks* to be a uniform purple to us if every one of these boxes of $n \times n \times n$ compartments has a value of r/b that is between 0.999 and 1.001 (so that r and b are the same, to an accuracy of one tenth of a per cent). This may seem, at first consideration, to be a rather stringent requirement (having to apply to *every* individual $n \times n \times n$ compartment). But when n gets very large, we find that the vast majority of the ball arrangements *do* satisfy this condition! We should also bear in mind that when considering molecules in a can of paint, the number of them will be staggeringly large, by ordinary standards. For example, there could well be something like 10^{24} molecules in an ordinary can of paint, so taking $N=10^8$ would not be at all unreasonable. Also, as will be clear when we consider that colours look perfectly good in digitally displayed photographs with a pixel size of only 10^{-2} cm, taking a value of $k=10^3$ is also very reasonable, in this model. From this, we find that, with these numbers ($N=10^8$ and $k=10^3$, so $n=10^5$) there are around $10^{23\,570000\,000000\,000000\,000000}$ different arrangements of the entire collection of $\frac{1}{2}N^3$ red balls and $\frac{1}{2}N^3$ blue balls that give the appearance of a uniform purple. There are only a mere $10^{46\,500000\,000000}$ different arrangements which give the original configuration in which the blue is entirely at the top and the red entirely at the bottom. Thus, for balls distributed entirely at random, the probability of finding uniform purple is a virtual certainty, whereas the probability of finding all the blue ones at the top is something like $10^{-23\,570000\,000000\,000000\,000000}$ (and this figure is not substantially changed if we do not require 'all' the blue balls to be initially at the top but, say, only 99.9% of them to be at the top).

We are to think of the 'entropy' to be something like a measure of these probabilities or, rather, of these different numbers of arrangements that give the same 'overall appearance'. Actually, to use these numbers directly would give an exceedingly unruly measure, owing to their vast differences in size. It is fortunate, therefore, that there are good theoretical reasons for taking the (natural) *logarithm* of these numbers as a more appropriate 'entropy' measure. For those readers who are not very familiar with the notion of a logarithm (especially a 'natural' logarithm), let us phrase things in terms of the logarithm taken to the base 10—referred to here as '$\log_{10}$' (rather than the natural logarithm, used later, which I refer to simply as 'log'). To understand $\log_{10}$, the basic thing to remember is that

$$\log_{10} 1=0,\ \log_{10} 10=1,\ \log_{10} 100=2,\ \log_{10} 1000=3,\ \log_{10} 10\,000=4, \text{ etc.}$$

That is, to obtain the $\log_{10}$ of a power of 10, we simply count the number of 0s. For a (positive) whole number that is not a power of 10, we can generalize this to say that the integral part (i.e. the number before the decimal point) of its $\log_{10}$ is obtained by counting the total number of digits and subtracting 1, e.g. (with the integral part printed in bold type)

$$\log_{10} 2=\mathbf{0}.301\,029\,995\,66\ldots$$
$$\log_{10} 53=\mathbf{1}.724\,275\,869\,60\ldots$$
$$\log_{10} 9140=\mathbf{3}.960\,946\,195\,73\ldots$$

etc., so in each case the number in bold type is just one less than the number of digits in the number whose $\log_{10}$ is being taken. The most important property of $\log_{10}$ (or of log) is that it *converts multiplication to addition*; that is:

$$\log_{10}(ab)=\log_{10} a+\log_{10} b.$$

(In the case when a and b are both powers of 10, this is obvious from the above, since multiplying $a=10^A$ by $b=10^B$ gives us $ab=10^{A+B}$.)

The significance of the above displayed relation to the use of the logarithm in the notion of entropy is that we want the entropy of a system which consists of two separate and completely independent components to be what we get by simply *adding* the entropies of the individual parts. We say that, in this sense, the entropy concept is *additive*. Indeed, if the first component can come about in P different ways and the second component in Q different ways, then there will be the product PQ of different ways in which the entire system—consisting of both components together—can come about (since to each of the P arrangements giving the first component there will be exactly Q arrangements giving the second). Thus, by defining the entropy of the state of any system to be proportional to the *logarithm* of the number of different ways that that state can come about, we ensure that this additivity property, for independent systems, will indeed be satisfied.

I have, however, been a bit vague, as yet, about what I mean by this 'number of ways in which the state of a system can come about'. In the first place, when we model the locations of molecules (in a can of paint, say), we would normally not consider it realistic to have discrete compartments, since in Newtonian theory there would, in full detail, be an infinite number of different possible locations for each molecule rather than just a finite number. In addition, each individual molecule might be of some asymmetrical shape, so that it could be oriented in space in different ways. Or it might have other kinds of internal degrees of freedom, such as distortions of its shape, which would have to be correspondingly taken into account. Each such orientation or distortion would have to count as a different configuration of the system. We can deal with all these points by considering what is known as the *configuration* space of a system, which I next describe.

For a system of d degrees of freedom, the configuration space would be a d-dimensional space. For example, if the system consisted of q point particles $p_1,p_2,\ldots,p_q$ (each without any internal degrees of freedom), then the configuration space would have $3q$ dimensions. This is because each individual particle requires just three coordinates to determine its position, so there are $3q$ coordinates overall, whereby a single point P of configuration space defines the locations of all of $p_1,p_2,\ldots,p_q$ together (see Fig. 1.4). In more complicated situations, where there are internal degrees of freedom as above, we would have more degrees of freedom for each such particle, but the general idea is the same. Of course, I am not expecting the reader to be able to 'visualize' what is going on in a space of such a high number of dimensions. This will not be necessary. We shall get a good enough idea if we just imagine things going on in a space of just 2 dimensions (such as a region drawn on a piece of paper) or of some region in ordinary 3-dimensional space, provided that we always bear in mind that such visualizations will inevitably be limited in certain ways, some of which we shall be coming to shortly. And of course we should always keep in mind that such spaces are purely abstract mathematical ones which should not be confused with the 3-dimensional *physical* space or 4-dimensional *physical* space-time of our ordinary experiences.

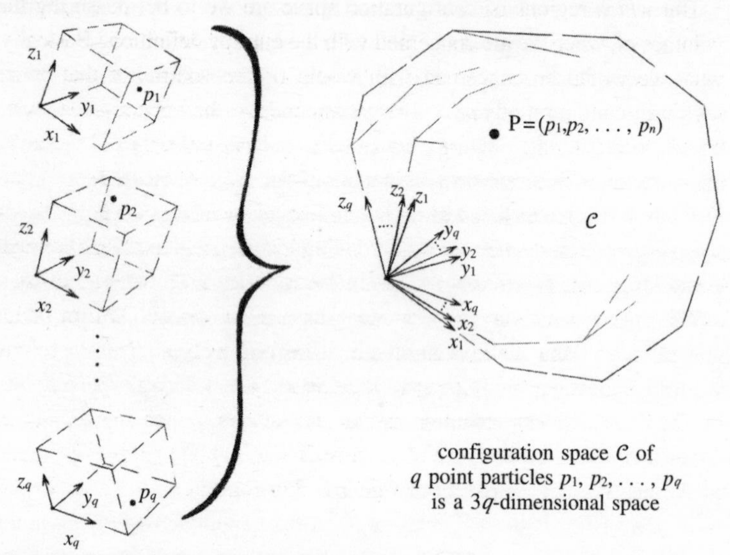

Fig. 1.4 Configuration space $\mathcal{C}$ of q point particles $p_1, p_2, \ldots, p_q$ is a $3q$-dimensional space.

There is a further point that needs clarification, in our attempts at a definition of entropy, and this is the issue of what exactly we are trying to count. In the case of our finite model, we had finite numbers of different arrangements for the red and blue balls. But now we have an infinite number of arrangements (since the particle locations require continuous parameters), and this leads us to consider high-dimensional *volumes* in configuration space, to provide us with an appropriate measure of *size*, instead of just counting discrete things.

To get an idea of what is meant by a 'volume' in a high-dimensional space, it is a good idea first to think of lower dimensions. The 'volume-measure' for a region of 2-dimensional curved surface, for example, would be simply the measure of surface *area* of that region. In the case of a 1-dimensional space, we are thinking simply of the *length* along some portion of a curve. In an n-dimensional configuration space, we would be thinking in terms of some n-dimensional analogue of the volume of an ordinary 3-volume region.

But *which* regions of configuration space are we to be measuring the volumes of, when we are concerned with the entropy definition? Basically, what we would be concerned with would be the volume of that entire region in configuration space that corresponds to the collection of states which 'look the same' as the particular state under consideration. Of course, 'look the same' is a rather vague phrase. What is really meant here is that we have some reasonably exhaustive collection of *macroscopic parameters* which measure such things as density distribution, colour, chemical composition, but we would not be concerned with such detailed matters as the precise locations of every atom that constitutes the system under consideration. This dividing up of the configuration space $\mathcal{C}$ into regions that 'look the same' in this sense is referred to as a 'coarse graining' of $\mathcal{C}$. Thus, each 'coarse-graining region' consists of points that represent states that would be considered to be indistinguishable from each other, by means of macroscopic measurements. See Fig. 1.5.

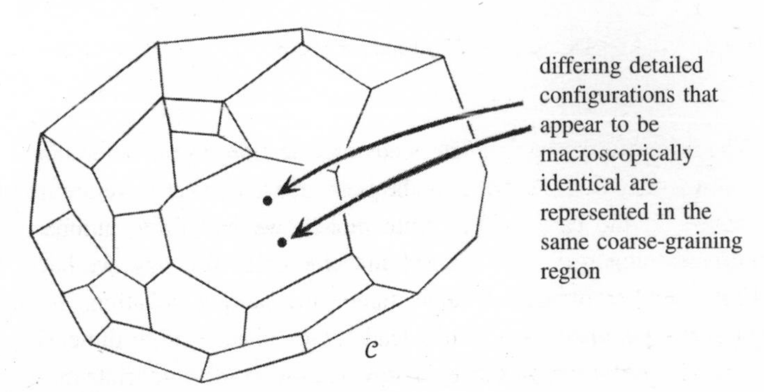

Fig. 1.5 A coarse-graining of $\mathcal{C}$.

Of course, what is meant by a 'macroscopic' measurement, is still rather vague, but we are looking for some kind of analogue of the 'hue' notion that we were concerned with above in our simplified finite model for the can of paint. There is admittedly some vagueness in such a 'coarse-graining' notion, but it is the *volume* of such a region in configuration space—or, rather, the *logarithm* of the volume of such a coarse-graining region—that we are concerned with in the definition of entropy. Yes, this is still a bit vague, but it is remarkable how robust the entropy notion turns out to be, nevertheless, mainly due to the absolutely stupendous ratios of volumes that the coarse-graining volumes turn out to have.

1.3 Phase space, and Boltzmann's definition of entropy

We are still not finished with the definition of entropy, however, for what has been said up to this point only *half* addresses the issue. We can see an inadequacy in our description so far by considering a slightly different example. Rather than having a can of red and blue paint, we might consider a bottle which is half filled with water and half with olive oil. We can stir it as much as we like, and also shake the bottle vigorously. But in a few moments, the olive oil and the water will separate out, and we soon have just olive oil at the top half of the bottle and water at the bottom half. The entropy has been increasing all the time throughout the separation process, nevertheless. The new point that arises here is that there is a strong mutual attraction between the molecules of olive oil which causes them to aggregate, thereby expelling the water. The notion of mere configuration space is not adequate to account for the entropy increase in this kind of situation, as we really need to take into account the *motions* of the individual particles/molecules, not just of their locations. Their motions will be necessary for us, in any case, so that the future evolution of the state is determined, according to the Newtonian laws that we are assuming to be operative here. In the case of the molecules in the olive oil, their strong mutual attraction causes their velocities to increase (in vigorous orbital motions about one another) as they get closer together, and it is the 'motion' part of the relevant space which provides the needed extra volume (and therefore extra entropy) for the situations where the olive oil is collected together.

The space that we need, in place of the configuration space $\mathcal{C}$ described above, is what is called *phase space*. The phase space $\mathcal{P}$ has *twice* as many dimensions (!) as $\mathcal{C}$, and each position coordinate for each constituent particle (or molecule) must have a corresponding 'motion' coordinate in addition to that position coordinate (see Fig. 1.6). We might imagine that the appropriate such coordinate would be a measure of *velocity* (or angular velocity, in the case of angular coordinates describing orientation in space). However, it turns out (because of deep connections with the formalism of *Hamiltonian theory*[1.1]) that it is the *momentum* (or angular momentum, in the case of angular coordinates) that we shall require in order to describe the motion. In most familiar situations, all we need to know about this 'momentum' notion is that it is the *mass times the velocity* (as already mentioned in §1.1). Now the (instantaneous) motions, as well as the positions, of all the particles composing our system are encoded in the location of a single point p in $\mathcal{P}$. We say that the *state* of our system is described by the location of p within $\mathcal{P}$.

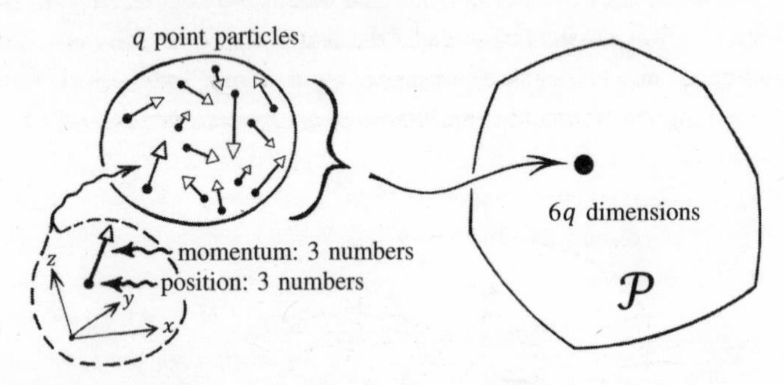

Fig. 1.6 The phase space $\mathcal{P}$ has twice as many dimensions as $\mathcal{C}$.

For the dynamical laws that we are considering, governing the behaviour of our system, we may as well take them to be Newton's laws of motion, but we can also treat more general situations (such as with the continuous fields of Maxwell's electrodynamics; see §2.6, §3.1, §3.2, and Appendix A1), which also come under the broad Hamiltonian framework (referred to above). These laws are *deterministic* in the sense that the state of our system at any one time completely determines the state at any other time, whether earlier or later. To put things another way, we can describe the dynamical evolution of our system, according to these laws as a point p which moves along a curve—called an *evolution curve*—in the phase space $\mathcal{P}$. This evolution curve represents the *unique* evolution of the entire system according to the dynamical laws, starting from the initial state, which we can represent by some particular point p_0 in the phase space $\mathcal{P}$. (See Fig. 1.7.) In fact, the whole phase space $\mathcal{P}$ will be filled up (technically *foliated*) by such evolution curves (rather like a bale of straw), where every point of $\mathcal{P}$ will lie on some particular evolution curve. We must think of this curve as being *oriented*—which means that we must assign a *direction* to the curve, and we can do this by putting an arrow on it. The evolution of our system, according to the dynamical laws, is described by a moving point p, which travels along the evolution curve—in this case starting from the particular point p_0—and moves in the direction in which the arrow points. This provides us with the future evolution of the particular state of the system represented by p. Following the evolution curve in the direction away from p_0 in the opposite direction to the arrow gives the time-reverse of the evolution, this telling us how the state represented by p_0 would have arisen from states in the past. Again, this evolution would be *unique*, according to the dynamical laws.

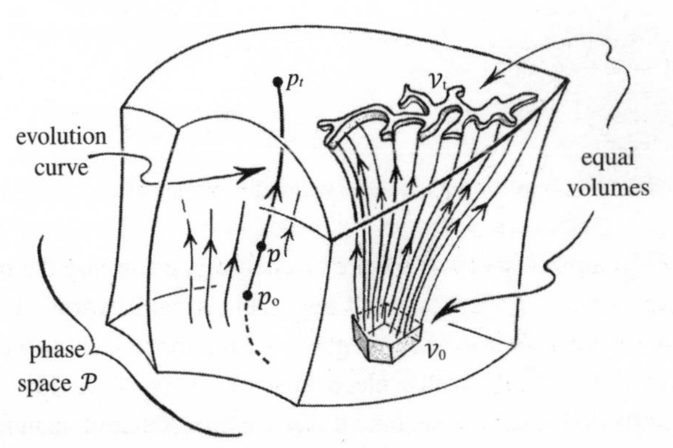

Fig. 1.7 Point p moves along an evolution curve in the phase space $\mathcal{P}$.

One important feature of phase space is that, since the advent of quantum mechanics, we find that it has a *natural measure*, so that we can take *volumes* in phase space to be, essentially, just dimensionless numbers. This is important, because Boltzmann's entropy definition, that we shall come to shortly, is given in terms of phase-space volumes, so we need to be able to compare high-dimensional volume measures with each other, where the dimensions may differ very greatly from one to another. This may seem strange from the point of view of ordinary classical (i.e. non-quantum) physics, since in ordinary terms we would think of the length of a curve (a 1-dimensional 'volume') as always having a smaller measure than the area of a surface (a 2-dimensional 'volume'), and a surface area as being of smaller measure than a 3-volume, etc. But the measures of phase-space volumes that quantum theory tells us to use are indeed just *numbers*, as measured in units of mass and distance that give us $\hbar = 1$, the quantity

$$\hbar = \frac{h}{2\pi}$$

being Dirac's version of Planck's constant (sometimes called the 'reduced' Planck's constant), where h is the original Planck's constant. In standard units, $\hbar$ has the extremely tiny value

$$\hbar = 1.05457\ldots \times 10^{-34} \text{ Joule seconds},$$

so the phase-space measures that we encounter in ordinary circumstances tend to have exceedingly large numerical values.

Thinking of these numbers as being just *integers* (whole numbers) gives a certain 'granularity' to phase space, and this provides the *discreteness* of the 'quanta' of quantum mechanics. But in most ordinary circumstances these numbers would be huge, so the granularity and discreteness is not noticed. An exception arises with the Planck black-body spectrum that we shall be coming to in §2.2 (see Fig. 2.6 and note 1.2), this being the observed phenomenon that Max Planck's theoretical analysis explained, in 1900, thereby launching the quantum revolution. Here one must consider an equilibrium situation simultaneously involving different numbers of photons, and therefore phase spaces of different dimensions. The proper discussion of such matters is outside the scope of this book,[1.3] but we shall return to the basics of quantum theory in §3.4.

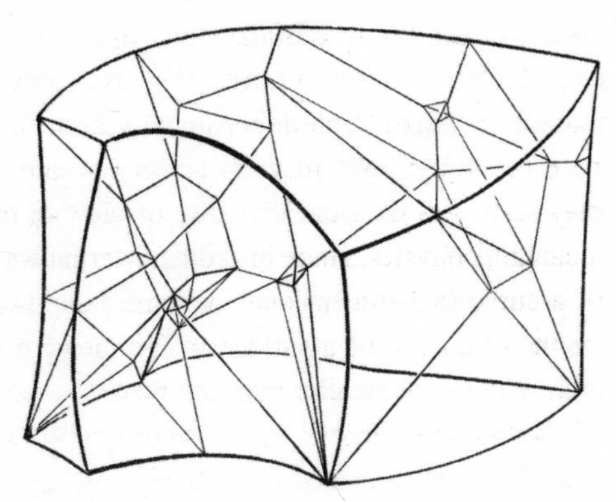

Fig. 1.8 Impression of coarse-graining in higher dimensions.

Now that we have the notion of the phase space of a system, we shall need to understand how the Second Law operates in relation to it. As with our discussion of configuration space, above, this will require us to provide a *coarse-graining* of $\mathcal{P}$, where two points belonging to the same coarse-graining region would be deemed to be 'indistinguishable' with regard to macroscopic parameters (such as the temperature, pressure, density, direction and magnitude of flow of a fluid, colour, chemical composition, etc.). The definition of the *entropy* S of a state represented by a point p in $\mathcal{P}$ is now provided by the remarkable Boltzmann formula

$$S = k' \log_{10} V,$$

where V is the volume of the coarse-graining region containing p. The quantity k' is a small constant – which would have been Boltzmann's constant had I chosen to use a natural logarithm, 'log'. It is given by $k' = k \log 10$ ($\log 10 = 2.302\,585\ldots$), where k is indeed Boltzmann's constant, and k takes the tiny value

$$k = 1.3805\ldots \times 10^{-23} \text{ Joules/degree Kelvin},$$

so $k' = 3.179\ldots \times 10^{-23}\,\text{JK}^{-1}$ (see Fig. 1.8). In fact, to be consistent with the definitions normally used by physicists, I shall henceforth revert to *natural* logarithms, and write Boltzmann's entropy formula as

$$S=k\log V,$$

where $\log V=2.302585\ldots\times\log_{10}V$.

Before we move on, in §1.4, to explore the reasonableness and implications of this elegant definition, and how it relates to the Second Law, we should appreciate one particular issue that is very nicely addressed by it. Sometimes people (quite correctly) point out that the lowness of the entropy of a state is not really a good measure of the state's 'specialness'. If we again consider the situation of the falling egg that was introduced in §1.1, we note that the relatively high-entropy state that is arrived at when the egg has become a mess on the floor is still an extraordinarily special one. It is special because there are some very particular correlations between the motions of the particles constituting that apparent 'mess'—of such a nature that if we *reverse* them all, then that mess will have the remarkable property that it will rapidly resolve itself into a perfectly completed egg that projects itself upwards so as to perch itself delicately on the table above. This, indeed, is a very special state, no less special than was the relatively low-entropy configuration of the egg up on the table in the first place. But, 'special' as that state consisting of a mess on the floor undoubtedly was, it was *not* special in the particular way that we refer to as 'low entropy'. Lowness of entropy refers to *manifest* speciality, which is seen in special values of the macroscopic parameters. Subtle correlations between particle motions are neither here nor there when it comes to the entropy that is to be assigned to the state of a system.

We see that although *some* states of relatively high entropy (such as the time-reversed smashed egg just considered) can evolve to states of *lower* entropy, in contradiction with the Second Law, these would represent a very tiny minority of the possibilities. It may be said that this is the 'whole point' of the notion of entropy and of the Second Law. Boltzmann's definition of entropy in terms of the notion of coarse graining deals with this matter of the kind of 'specialness' that is demanded by low entropy in a very natural and appropriate way.

One further point is worth making here. There is a key mathematical theorem known as *Liouville's theorem*, which asserts that, for the normal type of classical dynamical system considered by physicists (the standard *Hamiltonian* systems referred to above), the time-evolution preserves *volumes* in phase space. This is illustrated in the right-hand part of Fig. 1.7, where we see that if a region of $\mathcal{V}_0$, of volume V, in phase space $\mathcal{P}$, is carried by the evolution curves to a region $\mathcal{V}_t$, after time t, then we find that $\mathcal{V}_t$ has the same volume V as does $\mathcal{V}_0$. This does not contradict the Second Law, however, because the coarse-graining regions are not preserved by the evolution. If the initial region $\mathcal{V}_0$ happened to be a coarse-graining region, then $\mathcal{V}_t$ would be likely to be a sprawling messy volume spreading out through a much larger coarse-graining region, or perhaps several such regions, at the later time t.

To end this section, it will be appropriate to return to the important matter of the use of a *logarithm* in Boltzmann's formula, following up an issue that was briefly addressed in §1.2. The matter will have particular importance for us later, most especially in §3.4. Suppose that we are contemplating the physics taking place in our local laboratory, and we wish to consider the definition of the entropy of some structures involved in an experiment that we are performing. What would we consider to be Boltzmann's definition of entropy relevant to our experiment? We would take into account all degrees of freedom of concern, in the laboratory, and use these to define some phase space $\mathcal{P}$. Within $\mathcal{P}$ would be the relevant coarse-graining region $\mathcal{V}$ of volume V, giving us our Boltzmann entropy $k\log V$.

However, we might choose to consider our laboratory as part of a far larger system, let us say the rest of the entire Milky Way galaxy within which we reside, where there are enormously many more degrees of freedom. By including all these degrees of freedom, we find that our phase space will now be enormously larger than before. Moreover, the coarse-graining region pertinent to our calculation of entropies within our laboratory will now also be enormously larger than before, because it can involve all the degrees of freedom present in the entire galaxy, not just those relevant to the contents of the laboratory. This is natural, however, because the entropy value is now that which applies to the galaxy as a whole, the entropy involved in our experiment being only a small part of this.

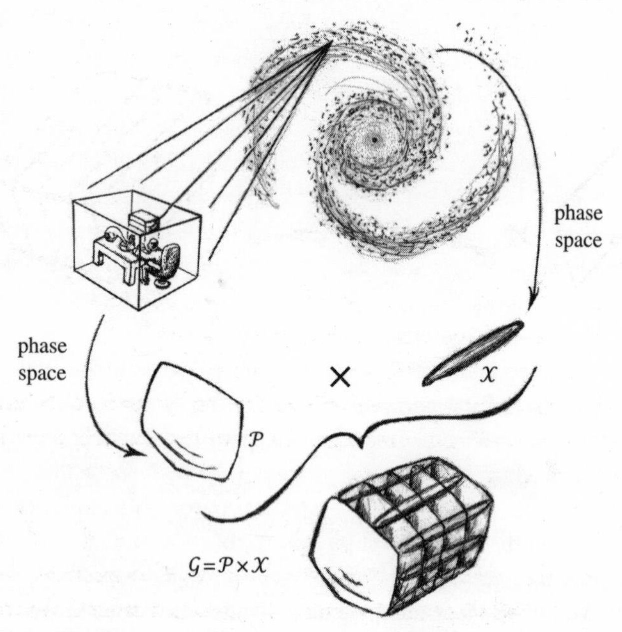

Fig. 1.9 The phase space considered by the experimenter is but a small factor of that which involves all the external degrees of freedom in the galaxy.

The parameters defining the *external* degrees of freedom (those determining the state of the galaxy *except* for those defining the state within the laboratory) provide us with a huge 'external' phase space $\mathcal{X}$, and there will be a coarse-graining region $\mathcal{W}$ within $\mathcal{X}$ that characterizes the state of the galaxy external to the laboratory. See Fig. 1.9. The phase space $\mathcal{G}$ for the entire galaxy will be defined by the complete set of parameters, both external (providing the space $\mathcal{X}$) and internal (providing the space $\mathcal{P}$). The space $\mathcal{G}$ is called, by mathematicians, the *product space*[1.4] of $\mathcal{P}$ with $\mathcal{X}$, written

$$\mathcal{G}=\mathcal{P}\times\mathcal{X},$$

and its dimension will be the *sum* of the dimensions of $\mathcal{P}$ and of $\mathcal{X}$ (because its coordinates are those of $\mathcal{P}$ followed by those of $\mathcal{X}$). Figure 1.10 illustrates the idea of a product space, where $\mathcal{P}$ is a plane and $\mathcal{X}$ is a line.

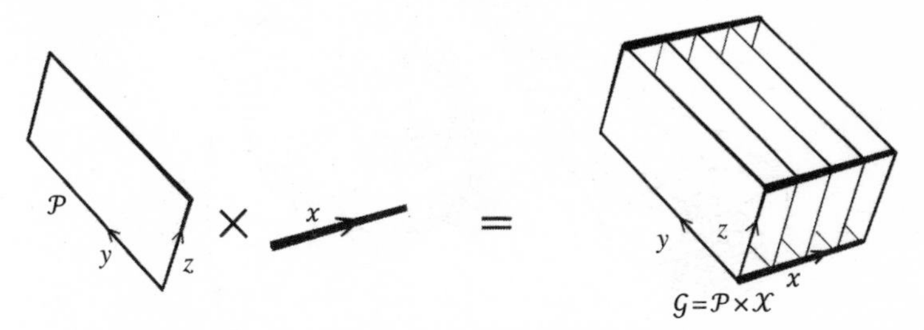

Fig. 1.10 Product space where $\mathcal{P}$ is a plane and $\mathcal{X}$ is a line.

If we take the external degrees of freedom to be completely independent of the internal ones, then the relevant coarse-graining region in $\mathcal{G}$ will be the product

$$\mathcal{V}\times\mathcal{W}$$

of the coarse-graining regions $\mathcal{V}$ in $\mathcal{P}$ and $\mathcal{W}$ in $\mathcal{X}$, respectively (see Fig.1.11). Moreover, the volume element in a product space is taken to be the product of the volume elements in each of the constituent spaces; consequently the volume of the coarse-graining region $\mathcal{V}\times\mathcal{W}$ in $\mathcal{G}$ will be the product VW of the volume V of the coarse-graining region $\mathcal{V}$ in $\mathcal{P}$ with the volume W of the coarse-graining region $\mathcal{W}$ in $\mathcal{X}$. Hence, by the 'product-to-sum' property of the logarithm, the Boltzmann entropy we obtain is

$$k \log (VW)=k \log V+k \log W,$$

which is the *sum* of the entropy within the laboratory and the entropy external to the laboratory. This just tells us that entropies of independent systems just *add* together, showing that an entropy value is something that can be assigned to any *part* of a physical system that is independent of the rest of the system.

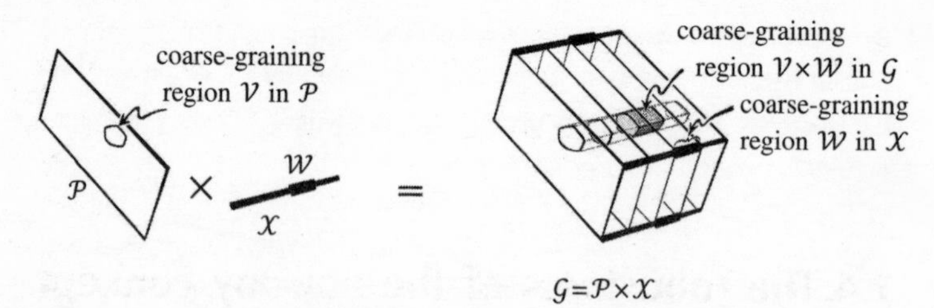

Fig. 1.11 Coarse-graining region in the product space as a product of coarse-graining regions in the factors.

In the situation considered here for which $\mathcal{P}$ refers to the degrees of freedom relevant to the laboratory and $\mathcal{X}$ to those relevant to the external galaxy (assumed independent of each other), we find that the entropy value $k\log V$ that the experimenter would assign to the experiment being performed, if the external degrees of freedom are being ignored, would differ from the entropy value $k\log(VW)$ that would result if these external degrees of freedom are also taken into consideration, simply by the entropy value $k\log W$ that would be assigned to all the external galactic degrees of freedom. This external part plays no role for the experimenter and can therefore be safely ignored for studying the role of the Second Law within the laboratory itself. However, when in §3.4 we come to consider the entropy balance of the universe *as a whole* and, most particularly, the contribution due to black holes, we shall find that these matters cannot be ignored, and will acquire a fundamental significance for us!

1.4 The robustness of the entropy concept

Matters concerning the entropy of the entire cosmos can be left aside for the time being. For the moment, we can just appreciate the value of Boltzmann's formula, for it provides us with an excellent notion of what the entropy of a physical system should be actually defined to be. Boltzmann put forward this definition in 1875, and it represented an enormous advance on what had gone before,[1.5] so that it now becomes possible to apply the entropy concept in completely general situations, where no assumption need be made, such as the system in question having to be in some kind of stationary state. There are, nevertheless, still certain aspects of vagueness in this definition, associated, primarily, with the notion of what is to be meant by a 'macroscopic parameter'. We might, for example, imagine that it will become possible, in the future, to measure many detailed aspects of the state of a fluid, where these would today be considered to be 'unmeasurable'. Rather than being merely able to determine, say, what the pressure, density, temperature, and overall fluid velocity might be at various locations of the fluid, it might become possible in the future to ascertain the motions of the fluid molecules in a great deal more detail, where we might perhaps be able even to measure the motions of specific molecules in the fluid. Accordingly, the coarse-graining of the phase space would then have to be taken rather more finely that it had been before. Consequently, the entropy of a particular state of the fluid would be considered to be somewhat smaller, as judged in the light of this newer technology, than it would have been previously.

Some scientists have argued[1.6] that the use of such technology to ascertain finer details of a system in this way would always entail an entropy increase in the *measuring apparatus* which more than compensates the effective entropy reduction that would be ascertained to be taking place in the system under examination, by virtue of the detailed measurements. Accordingly, such detailed measurement of a system would still result in an increase in the entropy overall. This is a very reasonable point, but even if we take it into account, there is still a muddying of the Boltzmann entropy definition, as the lack of objectivity in what constitutes a 'macroscopic parameter' for the system as a whole is hardly clarified by such considerations.

An extreme example of this sort of thing was envisaged by the great nineteenth-century mathematical physicist James Clark Maxwell (whose equations for electromagnetism have been referred to earlier; §1.1, §1.3). Maxwell imagined a tiny 'demon' able to direct individual gas molecules one way or another, by opening or closing a little door, thereby enabling the Second Law, as applied to the gas itself, to be violated. Yet, to consider the entire system, including the body of Maxwell's demon itself, as a single physical entity, the actual sub-microscopic composition of the demon would have to be brought into the picture, and the Second Law should be restored once this is done.

In more realistic terms, we might imagine the demon to be replaced by some minute mechanical device, and we can argue that the Second Law still holds good for the entire structure. The issue of what constitutes a macroscopic parameter does not seem to me to be properly resolved by such considerations, however, and the very definition of entropy, for such a complicated system, remains somewhat enigmatic. It might indeed seem odd that an apparently well-defined physical quantity like the entropy of a fluid should be dependent upon the specific state of technology at the time!

Yet, it is remarkable how little the entropy values that would be assigned to a system are affected, in a general way, by developments in technology such as this. The entropy values that would be attributed to a system would, on the whole, change very little as a result of redrawing the boundaries of the coarse-graining regions in this kind of way, as might result from improved technology. We must indeed bear in mind that there is likely to be always some measure of subjectivity in the precise value of the entropy that one might assign to a system, on account of the precision that might be available in measuring devices at any one time, but we should not adopt the point of view that the entropy is not a physically useful concept for that kind of reason. In practice, this subjectivity will, in normal circumstances, amount to a very small factor. The reason for this is that the coarse-graining regions tend to have volumes that are absolutely stupendously different from one another, and the detailed redrawing of their boundaries will normally make no discernible difference to the entropy values that are assigned.

To get some feeling for this, let us return to our simplified description of the mixture of red and blue paint, where we modelled this by considering 10^{24} compartments, occupied by equal total numbers of red and blue balls. There, we considered the colour at the various locations to be purple if the ratio of blue balls in a $10^5 \times 10^5 \times 10^5$ cubical crate lay in the range 0.999 to 1.001. Suppose that, instead, by the use of finer precision instruments, we are able to judge the red/blue ratio of the balls at a much finer scale than before, and much more precisely. Let us suppose that the mixture is now judged to be uniform only if the ratio of red balls to blue balls is between 0.9999 and 1.0001 (so that the numbers of red and blue balls are now to be equal to an accuracy of one hundredth of a per cent), which is ten times as precise as we had demanded before, and that the region examined need now only be one half of the dimension—and therefore one eighth of the volume—that we had had to examine before in order to determine the hue. Despite this very considerably increased precision, we find that the 'entropy' we must assign to the 'uniformly purple' state ('entropy' in the sense of the log of the number of states that now satisfy this condition) hardly changes from the value that we had previously. Consequently, our 'improved technology' makes effectively no difference to the sort of entropy values that we get in this kind of situation.

This is only a 'toy model', however (and a toy model of configuration space rather than phase space) but it serves to emphasize the fact that such changes in the precision of our 'macroscopic parameters' in defining the 'coarse-graining regions' tend not to make much difference to the entropy values that are assigned. The basic reason for this entropy robustness is simply the enormous size of the coarse-graining regions that we encounter and, more particularly, of the vastness of the ratios of the sizes of different such regions. To take a more realistic situation, we might consider the entropy increase that is involved in the commonplace action of taking a bath! For simplicity, I shall not attempt to place an estimate on the not inconsiderable raising of entropy that occurs in the actual cleansing process(!), but I shall concentrate only on what is involved in the mixing together of the water that emerges from the hot and cold taps (either in the bath itself or in the interior of a mixer tap which might be attached to the bath). It would be not unreasonable to suppose that the hot water emerges at a temperature of around 50 °C and the cold, around 10 °C, where the total volume of water that finds itself in the bath is being taken to be 150 litres (made half from the hot water and half from the cold). The entropy increase turns out to be about 21407 J/K, which amounts to our point in phase space moving from one coarse-graining region to another that is about 10^{27} times larger! No reasonable-looking change in precisely where the boundary of the coarse-graining regions are to be drawn would make any significant impression on numbers of this scale.

There is another related issue that should be mentioned here. I have phrased things as though the coarse-graining regions are well defined, with definite boundaries, whereas strictly speaking this would not be the case, no matter what plausible family of 'macroscopic parameters' might be adopted. Indeed, wherever the boundary of a coarse-graining region might be drawn, if we consider two very close points in the phase space, one on either side of the boundary, the two would represent states that are almost identical, and therefore macroscopically identical; yet they have been deemed to be 'macroscopically distinguishable' by virtue of their belonging to different coarse-graining regions![1.7] We can resolve this problem by asking that there be a region of 'fuzziness' at the boundaries separating one coarse-graining region from the next and, as with the issue of subjectivity about what precisely is to qualify as a 'macroscopic parameter', we may choose simply not to care what we do with the phase-space points lying within this 'fuzzy boundary' (see Fig. 1.12). It is reasonable to consider that such points occupy a very tiny phase-space volume in comparison with the vast interiors of these coarse-graining regions. For this reason, whether we consider a point close to the boundary to belong to one region or to the other will be a matter of little concern, as it makes effectively no difference to the value of the entropy that would normally be assigned to a state. Again, we find that the notion of the entropy of a system is a very robust one—despite the lack of complete firmness in its definition—owing to the very vastness of the coarse-graining regions, and of the enormous imbalance between their sizes.

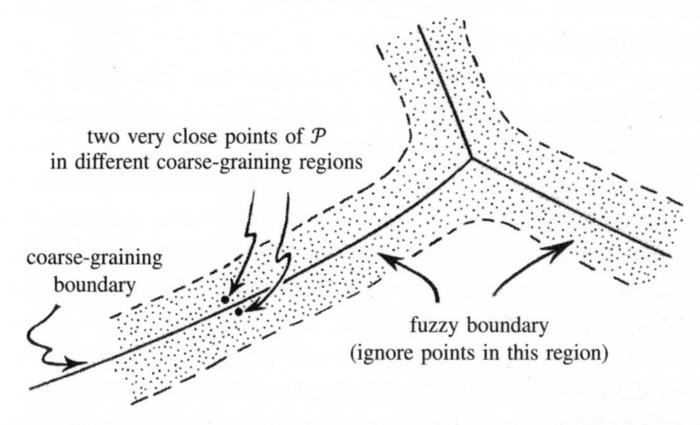

Fig. 1.12 'Fuzziness' at the boundaries separating one coarse-graining region from the next.

All this having been said, it must however be pointed out that there are various particularly subtle situations where such crude notions of 'macroscopic indistinguishability' would *appear* to be inadequate, and even seem to give us quite wrong answers for the entropy! One such situation occurs with the phenomenon of *spin echo* (first noticed by Erwin Hahn in 1950) that is made use of in connection with nuclear magnetic resonance (NMR). According to this phenomenon, some material with an initial specific state of magnetization, with nuclear spins[1.8] closely aligned, can lose this magnetization under the influence of a varying external electromagnetic field, the nuclear spins then taking up a much more higgledy-piggledy-appearing configuration owing to a complicated collection of spin precessions occurring at different rates. But if the external field is then carefully reversed, the nuclear spins all return to their original states, so that, very strikingly, the specific original magnetization state is retrieved! As far as macroscopic measurements are concerned, it would appear that the entropy had increased in the transition to this intermediate stage (with the higgledy-piggledy nuclear spins)—consistently with the Second Law—but when the nuclear spins regain the order that they had lost in the intermediate stage, as a result of the application of the reversed external electromagnetic field, it would appear that the Second Law has been grossly violated, owing to an entropy *decrease* during this final process![1.9]

The fact is that even though the spin states would appear to be very disordered in the intermediate situation, there is actually a very precise 'hidden order' in the apparently higgledy-piggledy arrangement of spins, this order being revealed only when the pattern of external magnetic field movements is reversed. Something analogous occurs with a CD or DVD, where any ordinary crude 'macroscopic measurement' would be likely not to reveal the very considerable stored information on such a disc, whereas an appropriate playing device specifically designed to read the disc would have no trouble in revealing this stored information. To detect this hidden order, one needs 'measurements' of a much more sophisticated type than the 'ordinary' macroscopic measurements that would be adequate in most situations.

We do not really need to consider anything so technically sophisticated as the examination of tiny magnetic fields to find 'hidden order' of this general kind. Something essentially similar occurs with a much simpler-looking apparatus (see Fig. 1.13, and for further information Note 1.10). This consists of two cylindrical glass tubes, one of which fits very snugly inside the other, there being a very narrow space between the two. Some viscous fluid (e.g. glycerine) is inserted uniformly into this thin space between the two cylinders, and a handle is attached appropriately to the inner one, so that it can be rotated with respect to the outer one which remains fixed. Now the experiment is set up so that there is a thin straight line of bright red dye inserted in the fluid, parallel to the axis of the cylinder (see Fig.1.14). The handle is then turned around several times, the line of dye spreading as a consequence of this, until it is observed to be distributed uniformly around the cylinder so that no trace of its original concentration along a line is now seen, but the fluid acquires a very faint pinkish hue. By any reasonable-looking choice of 'macroscopic parameters' to ascertain the state of the dyed viscous fluid, the entropy would appear to have gone up, the dye being now uniformly spread over the fluid. (The situation might appear to be very similar to what happened with the stirred combination of red and blue paint that we considered in §1.2.) However, if the handle is now rotated in the reverse direction, by just the same number of turns as had been used before, we find, rather miraculously, that the line of red dye reappears, and becomes almost as clearly defined as it had been in the first place! If the entropy had indeed been raised in the initial winding, by the amount that had appeared to be the case, and if the entropy is considered to have returned to close to its original value after the rewinding, then we have a severe violation of the Second Law as a result of this rewinding process!

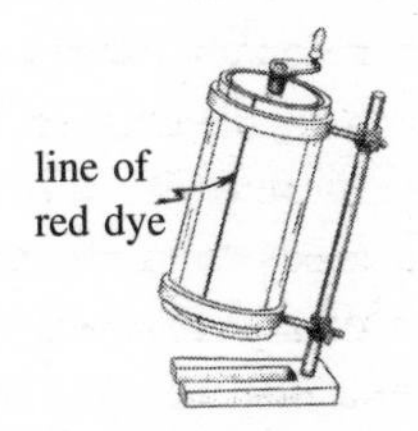

Fig. 1.13 Two snug-fitting glass tubes and viscous fluid between, with line of red dye.

In both these situations, it would be the common viewpoint that the Second Law has *not*, in actuality, been violated, but that in such situations the entropy definition has not been refined enough. In my opinion, there is a 'can of worms' here, if one demands that there should be a precise objective definition of physical entropy, applicable in *all* circumstances, with respect to which the Second Law is to be universally valid.

Fig. 1.14 The handle is turned several times spreading out the line of dye. The handle is then turned back the same number of times and the line reappears in apparent violation of the Second Law.

I do not see why one should demand that there always be a well-defined, physically precise notion of 'entropy', that is entirely objective and consequently 'out there' in Nature, in some absolute sense,[1.11] where this 'objective entropy' almost never decreases as time progresses. Must there always be an *actual* entropy concept that applies to the slightly tinted viscous fluid between the cylinders, or to the configurations of nuclear spins that had *appeared* to become totally disorganized, though retaining a precise 'memory' of the order that they had had before? I do not see that this need be the case. Entropy is clearly an extremely useful physical concept, but I do not see why it need be assigned a truly fundamental and objective role in physics. Indeed, it seems reasonable to me that the usefulness of the physical notion of entropy has its origin largely in the fact that, for systems that we tend to encounter in the *actual universe*, it turns out that the normal measures of 'macroscopic' quantities give rise to coarse-graining volumes that do in fact differ from one another by stupendously large factors. There is a profound issue, however, as to *why*, in the universe that we know, they should differ by such enormous factors. These enormous factors reveal a remarkable fact about our universe that *does* seem to be clearly objective and 'out there'—and we shall be coming to this shortly—despite the admittedly confusing issues of subjectivity that are involved in our concept of 'entropy', these serving merely to cloud the central mystery that underlies the profound usefulness of this remarkable physical notion.

1.5 The inexorable increase of entropy into the future

LET us try to get some understanding of why it is to be expected that the entropy should increase when a system evolves into the future, as the Second Law demands. Suppose we imagine that our system starts off in a state of reasonably low entropy—so that the point p, which is to move through phase space $\mathcal{P}$ thereby describing the time-evolution of the system, starts off at a point p_0 in a fairly small coarse-graining region $\mathcal{R}_0$ (see Fig.1.15). We must bear in mind that, as noted above, the various coarse-graining regions tend to differ in size by absolutely enormous factors. Also, the huge dimensionality of phase space $\mathcal{P}$ will tend to imply that there are likely to be vast numbers of coarse-graining volumes neighbouring any one particular region. (Our 2- or 3-dimensional images are rather misleading in this particular respect, but we see that the number of neighbours is going up with increasing dimension—typically six in the 2-dimension case and fourteen in 3; see Fig.1.16.) Thus, it will be exceedingly probable that the evolution curve described by p, as it leaves the coarse-graining region $\mathcal{R}_0$ of the starting point p_0 and enters the next coarse-graining region $\mathcal{R}_1$, will find that $\mathcal{R}_1$ has a hugely greater volume than $\mathcal{R}_0$—for to find, instead, an enormously smaller volume would seem a grossly unlikely action for the point p to take, as though p were to succeed, just by *chance*, in the proverbial search for a needle in a haystack, but here with an enormously more formidable task!

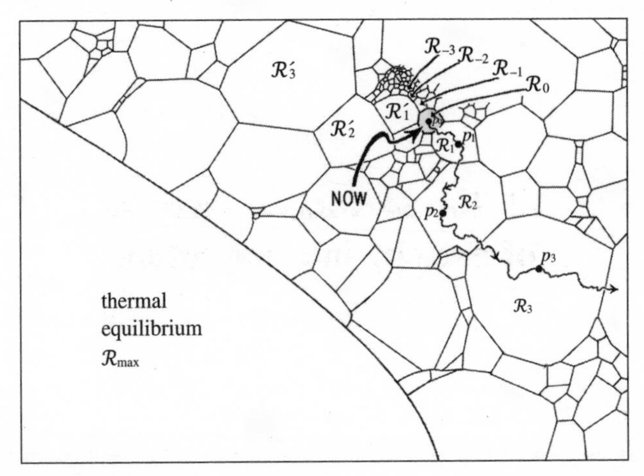

Fig. 1.15 The system starts off at a point of p_0 in a fairly small coarse-graining region $\mathcal{R}_0$.

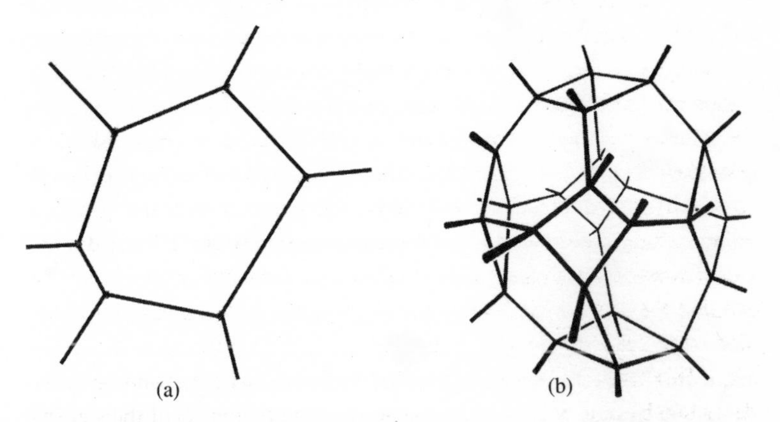

Fig. 1.16 As the dimension n increases, the typical number of neighbouring coarse-graining regions increases rapidly. (a) $n=2$ with typically 6 neighbours. (b) $n=3$ with typically 14 neighbours.

The *logarithm* of $\mathcal{R}_1$'s volume will also, consequently, be somewhat greater than the logarithm of of $\mathcal{R}_0$'s volume, though only a good deal more modestly greater than is provided by the increase in actual volume (see §1.2 above), so the entropy will have been increased just somewhat. Then, as p enters the next coarse-graining region, say $\mathcal{R}_2$, we are again highly likely to find that the volume of $\mathcal{R}_2$ is hugely greater than that of $\mathcal{R}_1$, so the entropy value will again grow somewhat. We next expect to find that p subsequently enters another region, say $\mathcal{R}_3$, even hugely larger than those it had been in before, and the entropy goes up again somewhat, and so on. Moreover, because of the vast increases in the volumes of these coarse-graining regions, once p has entered a larger region, we may regard it as a practical impossibility—i.e. as 'overwhelmingly unlikely'—that it will find a smaller one again, of the kind of far tinier sizes that provided the somewhat smaller entropy values that were encountered previously. Thus, as time marches forward, into the future, the entropy value must be expected to increase relentlessly, though far more modestly than do the actual volumes.

Of course, it is not strictly impossible that a smaller entropy value may be obtained in this way; it is merely that such occurrences of entropy reduction must be regarded as overwhelmingly unlikely. The entropy increase that we have obtained is simply the kind of trend that we must take to be the normal state of affairs where the evolution proceeds in a way that has no particular bias with regard to the coarse-graining of the phase space, and might as well be treated as though the track of p through phase space were essentially random, despite the fact that the evolution is actually governed by the well-defined and completely deterministic procedures of (say) Newtonian mechanics.

One might legitimately wonder why p does not simply *directly* enter $\mathcal{R}_{max}$, the (vastly) greatest coarse-graining region of all, rather than sequentially entering a succession of larger and larger coarse-graining regions as described above. Here, $\mathcal{R}_{max}$ refers to what is commonly called *thermal equilibrium*, where the volume of $\mathcal{R}_{max}$ would be likely to exceed the total of all the other coarse-graining regions put together. Indeed, it may be expected that p will *eventually* reach $\mathcal{R}_{max}$, and when it does so it will, for the most part, remain in this region, with merely the very occasional excursion into a smaller region (a thermal fluctuation). But the evolution curve must be regarded as describing a continuous evolution, where the state at one moment is not likely to differ greatly from the state a moment before. Accordingly, the coarse-graining volumes would be likely not to differ from their neighbours by such an enormous amount as would be represented by a direct leap to $\mathcal{R}_{max}$, despite the vast changes in coarse-graining volumes that the evolution curve would encounter. We would not expect that the entropy is likely to jump that discontinuously, but merely to pass fairly gradually to larger and larger values of the entropy.

This appears to be all pretty satisfactory, and might well lead us to believe that a gradual entropy increase into the future is a completely natural expectation which seems to be hardly in need of further deep deliberation—except perhaps for details of rigour that might be needed to satisfy the mathematical purist. The egg, referred to in the previous section, which starts, at the moment NOW, by being perched on the edge of the table, indeed has a likely entropy-increasing future evolution that would be consistent with its falling off the table and smashing on the ground. This is completely in accordance with the simple considerations of greatly increasing phase-space volumes, as indicated above.

However, let us pose another question, somewhat different from that of the expected future behaviour of the egg. Let us ask for the likely *past* behaviour of the egg. We want to know, instead: 'what is the most likely way for the egg to have found itself to be perched on the edge of the table in the first place?'

We can attempt to address this issue in just the same way as before, where we asked for the most likely future evolution of our system starting from NOW, but this time we are asking for the most probable *past* evolution of our system *leading up to* NOW. Our Newtonian laws work just as well in the past time-direction, and again give us a deterministic past-evolution. Thus, there is some evolution curve leading up to the point p_0, in the phase space $\mathcal{P}$, which describes this past-evolution, and represents the way that the egg happened to become poised at the edge of the table. To find this 'most probable' past history of our egg, we again examine the coarse-graining regions adjoining $\mathcal{R}_0$, and we again observe that there are vast differences in their sizes. Accordingly, there will be enormously more evolution curves ending at p_0 which enter $\mathcal{R}_0$ from a huge region like $\mathcal{R}_1$, whose volume greatly exceeds that of $\mathcal{R}_0$, than there will be that enter $\mathcal{R}_0$ from much smaller regions. Let us say that the evolution curve enters $\mathcal{R}_0$ from the region $\mathcal{R}_1'$, very much larger than $\mathcal{R}_0$. Prior to this, there would again be neighbouring regions of vastly differing sizes, and we again note that the enormous majority of evolutions entering $\mathcal{R}_1'$ would come from coarse-graining regions far larger than $\mathcal{R}_1'$. Accordingly, it appears that we may again suppose that our past-evolution curve entering $\mathcal{R}_1'$ comes from some region $\mathcal{R}_2'$, of vastly greater volume than $\mathcal{R}_1'$, and that likewise it entered $\mathcal{R}_2'$, from a region $\mathcal{R}_3'$, of even larger volume than $\mathcal{R}_2'$, and so on. See Fig. 1.15.

This is the conclusion that our reasoning seems to have led us to, but does it make sense? Such evolution curves would be hugely more numerous than the evolution curves leading up to p_0 from the succession of much *smaller* volumes, say $\ldots, \mathcal{R}_{-3}, \mathcal{R}_{-2}, \mathcal{R}_{-1}, \mathcal{R}_0$, which would be likely to have *actually* occurred, whose volumes would be greatly increasing from smaller volumes, in the direction of increasing time, as would be consistent with the Second Law. Rather than providing us with support for the Second Law, our line of reasoning now seems to have led us to a completely *wrong* answer, namely to expect continual gross *violations* of the Second Law in the past!

Our reasoning seems to have led us to expect, for example, that an exceedingly probable way that our egg originally found itself to be perched at the edge of the table was that it started as a mess of broken eggshell at the bottom of the table, mixed with yolk and albumen all churned up together and partly absorbed between the floorboards. This mess then spontaneously collected itself together, removing itself cleanly from the floor, with yolk and albumen thoroughly separating out and becoming completely enclosed by the miraculously self-assembling eggshell to provide a perfectly constructed egg, which propels itself from the ground at exactly the right speed so that it can become delicately perched on the edge of the table. Such behaviour would be of the kind that our above reasoning led to, with a 'probable' evolution curve successively passing through regions with volumes of greatly reducing size, like $\ldots, \mathcal{R}_3', \mathcal{R}_2', \mathcal{R}_1', \mathcal{R}_0$. But this would be grossly in *conflict* with what presumably actually happened, namely that some careless person placed the egg on the table, not realizing that it was in danger of rolling too close to the edge. *That* evolution would have been *consistent* with the Second Law, being represented in the phase space $\mathcal{P}$, by an evolution curve passing through the succession of greatly *increasing* volumes . . . , $\mathcal{R}_{-3}$, $\mathcal{R}_{-2}$, $\mathcal{R}_{-1}$, $\mathcal{R}_0$. When applied in the past time-direction, our argument has indeed given us an answer that is about as completely *wrong* as one could imagine.

1.6 Why is the past different?

Why has our reasoning gone so sadly astray—this being apparently just the same reasoning that seemed convincingly to lead us to expect that the Second Law, with overwhelming probability, must hold for the future evolution of an ordinary physical system? The trouble with the reasoning, as I have provided it, lies in the assumption that the evolution can be regarded as effectively 'random' in relation to the coarse-graining regions. Of course it is not really random, as noted above, since it is precisely determined by the dynamical (e.g. Newton's) laws. But we have taken it that there is no particular bias in this dynamical behaviour, in relation to these coarse-graining regions, and this supposition seemed to be fine for the future evolution. When we consider the past evolution, however, we find that this is manifestly not the case. There is a great deal of bias, for example, in the past-evolved behaviour of the egg, where it would appear to be guided inexorably—if viewed from a time-reversed perspective—from an original messy broken state, through exceptionally improbable actions albeit all in accordance with the dynamical laws, to the exceedingly improbable state of being balanced, complete and unbroken, at the edge of the table. If such behaviour were to be observed in *future*-directed behaviour it would be regarded as an impossible form of teleology or magic. Why do we regard such clearly focused behaviour as being perfectly acceptable if it is directed towards the past, whereas it would be rejected as scientifically unacceptable if directed into the future?

The answer—though hardly a 'physical explanation'—is simply that such 'past-teleology' is common experience, whereas 'future-teleology' is just something that we never seem to encounter. For it is just a *fact* of the observed universe that we do not encounter such 'future-teleology'; it is just observational fact that the Second Law holds good. In the universe we know, the dynamical laws appear not to be guided in any way to a future goal and can be regarded as being completely unconcerned with coarse-graining regions; whereas such 'guidance' of the evolution curve in past directions is utterly commonplace. If we examine the evolution curve in its past behaviour, it seems to be 'deliberately' seeking ever smaller and smaller coarse-graining regions. That we do not regard this as strange is simply a matter of it being such a familiar part of our everyday experience. The experience of an egg rolling off the edge of a table and smashing on the floor below is not regarded as strange, whereas a movie film of such an occurrence which is run in the reverse time-direction does indeed look extremely odd, and it represents something that in the ordinary time-direction is simply not part of our experience of the physical world. Such 'teleology' is perfectly acceptable if we are looking towards the past, but it is not a feature of our experience that it apply towards the future.

In fact we can understand this seeming past-teleology of behaviour if we simply suppose that the very origin of our universe was represented in phase space by a coarse-graining region of quite exceptional tininess, so that the initial state of the universe was one of particularly small entropy. So long as we may take it that the dynamical laws are such that there is an appropriate degree of continuity in the way that the entropy of the universe behaves, as noted above, then we need merely suppose that the universe's initial state—what we call the *Big Bang*—had, for some reason, an extraordinarily tiny entropy (a tininess which, as we shall be seeing in the next part, has a rather subtle character). The appropriate continuity of entropy would then imply a relatively gradual increase of the universe's entropy from then on (in the normal time-direction), giving us some kind of theoretical justification of the Second Law. So the key issue is indeed the specialness of the Big Bang, and the extraordinary minuteness of the initial coarse-graining region $\mathcal{B}$ that represents the nature of this special initial state.

The issue of the Big-Bang specialness is central to the arguments of this book. In §2.6 we shall be seeing how extraordinarily special the Big Bang must actually have been and we shall have to confront the very particular nature of this initial state. The underlying deep puzzles that this raises will later lead us into the strange line of thought that provides the distinctive underlying theme of this book. But just for the moment, we may simply take note of the fact that once we accept that such an extraordinarily special state did indeed originate the universe that we know, then the Second Law, in the form that we observe it, is a natural consequence. Provided that there is no corresponding low-entropy *ultimate* state of the universe, or some such, providing us with a teleological demand that the universe's evolution curve has to terminate in some other extraordinarily tiny 'future' region $\mathcal{F}$ in $\mathcal{P}$, then our reasoning for the increase of entropy in the future time-direction seems to be perfectly acceptable. It is the initial low-entropy constraint, demanding that the evolution curve originate within the extraordinarily tiny region $\mathcal{B}$ that gives us a theoretical basis for the Second Law that we actually experience in our universe.

A few points of clarification should however be addressed before we venture (in Part 2) into a more detailed examination of the Big-Bang state. In the first place, it has occasionally been argued that the existence of a Second Law holds no mystery, for our experience of the passage of time is dependent upon an increasing entropy as part of what constitutes our conscious feeling of the passage of time; so whatever time-direction we believe to be the 'future' must be that in which entropy increases. According to this argument, had the entropy been *decreasing* with respect to some time-parameter t, then our conscious feelings of temporal flow would project in the reverse direction, so that we would regard the small values of t to lie in what we think of as our 'future' and the large values in our 'past'. We would therefore take the parameter t to be the reverse of a normal time parameter, so that the entropy would still be increasing into what we experience as being the future. Thus, so the argument goes, our psychological experiences of the passage of time would always be such that the Second Law holds true, irrespective of the physical direction of the progression of entropy.

However, even leaving aside the very dubious nature of any such argument from our 'experience of time progression'—when we know almost nothing about what physical prerequisites might be required for 'conscious experience'—this argument misses the crucial point that the very usefulness of the notion of entropy depends upon our universe being enormously far from thermal equilibrium, so that coarse-graining regions that are far smaller than $\mathcal{R}_{\max}$ are involved in our common experience. In addition to this, the very fact that entropy is either uniformly increasing or uniformly decreasing depends upon the actuality of one or the other end (but not both ends) of the evolution curve in phase space being constrained to a very tiny coarse-graining region, and this is the case for only a very minute fraction of possible universe histories. It is the very tininess of the coarse-graining region $\mathcal{B}$ that our evolution curve appears to have encountered that needs explaining, and this issue is completely untouched by the aforementioned argument.

Sometimes the argument is made (perhaps in conjunction with the above) that the presence of a Second Law is an essential prerequisite for life, so that living beings like ourselves could only exist in a universe (or a universe epoch) in which the Second Law holds true, this law being a necessary ingredient of natural selection, etc. This is an example of 'anthropic reasoning' and I shall be returning briefly to this general issue in §3.2 (end) and §3.3. Whatever value this type of argument may have in other contexts, it is next to useless here. Again there is the very dubious aspect of such reasoning that we do not have a great deal more understanding of the physical requirements for life than we do for consciousness. But even apart from this, and even assuming that natural selection is indeed an essential prerequisite for life, and that it does require the Second Law, this still provides no explanation for the fact that the same Second Law operative here on Earth appears to hold everywhere in the observable universe to distances far beyond those of any relevance to local conditions, such as in galaxies thousands of millions of light years distant, and to times far earlier than the beginnings of life on Earth.

One further point to bear in mind is the following. If we do not *assume* the Second Law, or that the universe originated in some extraordinarily special initial state, or something else of this general nature, then we cannot use the 'improbability' of the existence of life as a premise for a derivation of a Second Law that is operative at times earlier than the present. No matter how curious and non-intuitive it may seem, the production of life would (if we do not make a prior *assumption* of the Second Law) be far *less* probable to come about by natural means—be it by natural selection or any by other seemingly 'natural' process—than by a 'miraculous' creation simply out of random collisions of the constituent particles! To see why this must be, we again examine our evolution curve in the phase space $\mathcal{P}$. If we consider the coarse-graining region $\mathcal{L}$ which represents our present-day Earth, teeming with life as it is, and we ask for the most probable way that this situation can have come about, then we again find that—as with our sequence of greatly decreasing coarse-graining regions . . . , $\mathcal{R}_3'$, $\mathcal{R}_2'$, $\mathcal{R}_1'$, $\mathcal{R}_0$ considered in §1.5 above—the 'most probable' way in which $\mathcal{L}$ would have been reached would have been via some corresponding sequence of coarse-graining regions . . ., $\mathcal{L}_3'$, $\mathcal{L}_2'$, $\mathcal{L}_1'$, $\mathcal{L}$ of greatly *decreasing* volume, representing some completely random-looking teleological assembly of life, completely at odds with what actually happened, this being violently in *disagreement* with the Second Law, rather than providing a demonstration of it. Accordingly, the mere existence of life provides, in itself, no argument whatever for the full validity of the Second Law.

There is a final point that should be addressed here, having to do with the *future*. I have argued that it is just a matter of observational fact that the Second Law, with its consequence of an enormous constraint on the initial state, actually holds true in our universe. It is again just a matter of observation that there appears *not* to be a corresponding constraint in the very remote future. But do we *really* know that the latter is actually the case? We do not really have much direct evidence of what, in detail, the *very* remote future will be like. (The evidence that we do have will be discussed in §3.1, §3.2, and §3.4.) We can certainly say that there is no indication available to us now, that hints that the entropy might eventually start to come down again, so that ultimately the very reverse of the Second Law might hold in the remote future. Yet, I do not see that we can absolutely rule out such a thing for the actual universe that we inhabit. Although the $\sim 1.4 \times 10^{10}$ years that have elapsed since the Big

Bang may seem to be a very long time (see §2.1), and no such reversed Second-Law effects have been observed, this time-span is as nothing when compared with what seems to be projected for the complete future time-span of the universe (as we shall be coming to address in §3.1)! In a universe constrained to have an evolution curve that *terminates* within some tiny region $\mathcal{F}$, its very late evolution must ultimately begin to experience curious correlations between particles that would eventually lead to the kind of teleological behaviour that would seem to us now to be as strange as the self-assembling egg described in §1.5 above.

There is no inconsistency with (say Newtonian) dynamics for the evolution curve of the universe, in its phase space $\mathcal{P}$, to be constrained so that it originates in one very tiny coarse-graining region $\mathcal{B}$ and *also* terminates in another one $\mathcal{F}$. There would be far fewer such curves than there would be those which merely originate in $\mathcal{B}$, but we must already be accustomed to the fact that those which merely originate in $\mathcal{B}$, as appears to be the case with the actual universe we inhabit, represent but an extremely minute proportion of the totality of possibilities. The situations in which the evolution curve is indeed constrained to have *both* end-points in very tiny regions would represent an even far tinier fraction of all the possibilities, but their logical status is not very different, and we cannot just rule them out. For such evolutions there would be a Second Law operating in the early stages of the universe, just as seems to be the case for the universe we know, but in the very late stages we should find that a *reverse* Second Law holds true, with entropy ultimately *decreasing* with time.

I do not myself regard as at all plausible this possibility that the Second Law might eventually reverse—and it will play no significant role in the main proposal that I shall be making in this book. Yet it should be made clear that while our experience provides no hint of such an ultimate reversal of the Second Law, such an eventuality is not intrinsically absurd. We must keep an open mind that exotic possibilities of this kind cannot necessarily be ruled out. In Part 3 of this book, I shall be making a different proposal, and an open mind will be helpful, also, for appreciating what I have to say. Yet the ideas are based on some remarkable facts about the universe, about which we can be reasonably certain. So let us start, in Part 2, with what we actually know about the Big Bang.

Part 2

The oddly special nature of the Big Bang

2.1 Our expanding universe

THE Big Bang: what do we believe actually happened? Is there clear observational evidence that a primordial explosion actually took place—from which the entire universe that we know appears to have originated? And, central to the issues raised in Part 1 is the question: how can such a wildly hot violent event represent a state of extraordinarily *tiny* entropy?

Initially, the main reason for believing in an explosive origin for the universe came from persuasive observations by the American astronomer Edwin Hubble that the universe is *expanding*. This was in 1929, although indications of this expansion had been previously noticed by Vesto Slipher in 1917. Hubble's observations demonstrated rather convincingly that distant galaxies are moving away from us with speeds that are basically proportional to their distances from us, so that if we extrapolate backwards, we come to the conclusion that everything would have come together at more or less the same time. That event would have constituted one stupendous explosion—what we now refer to as the 'Big Bang'—at which all matter appears to have had its ultimate origin. Subsequent observations, of which there are many, and detailed specific experiments (some of which I shall come to shortly), have confirmed and greatly strengthened Hubble's initial conclusions.

Hubble's considerations were based on the observations of a *red shift* in the spectral lines in the light emitted by distant galaxies. The term 'red shift' refers to the fact that the spectrum of frequencies emitted by different kinds of atoms in a distant galaxy is seen, on Earth, as being slightly shifted in the direction of red (Fig. 2.1), which is a uniform reduction in frequency consistent with an interpretation as the *Doppler shift*,[2.1] i.e. a reddening due to the observed object receding from us at considerable speed. The red shift is greater for galaxies that appear to be more distant, and the correlation with apparent distance turns out to be consistent with Hubble's picture of a spatially uniform expansion of the universe.

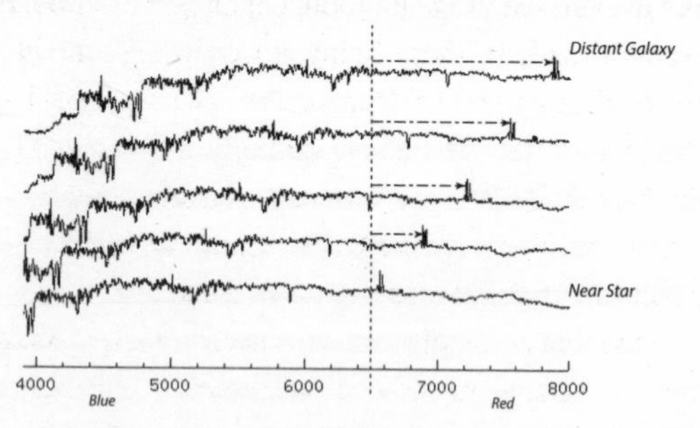

Fig. 2.1 The 'red shift' of the spectrum emitted by atoms in a distant galaxy is consistent with an interpretation as a Doppler shift.

Many refinements in the observations and their interpretation have occurred in the succeeding years, and it is fair to say that not only has Hubble's original contention been confirmed, in general terms, but recent work has given a fairly detailed view of how the expansion rate of the universe has evolved with time, providing us with a picture which is now pretty well generally accepted (although there are still some noteworthy dissenting voices,[2.2] when it comes to some of the detailed issues). In particular, a rather firm, generally agreed date of close to 1.37×10^{10} years ago, has been established for the moment when the matter of the universe would have to have been all together at its starting point—at what we indeed refer to as the 'Big Bang'.[2.3]

One should not think of the Big Bang as being localized at some particular region of space. The view that cosmologists take, in accordance with Einstein's perspective of general relativity, is that at the time that it occurred, the Big Bang encompassed the entire spatial spread of the universe, so it included the totality of all of physical *space*, not merely the material content of the universe. Accordingly space itself is taken to have been, in an appropriate sense, very tiny at the time. To understand such confusing matters, it is necessary to have some idea of how Einstein's curved-space-time general theory of relativity works. In §2.2, I shall have to address Einstein's theory in a fairly serious way, but for the moment, let us content ourselves with an analogy that is frequently used, namely that of a balloon which is being blown up. The universe, like the surface of the balloon, expands with time, but the whole of space expands with it, there being no central point in the universe from which it all expands. Of course the 3-dimensional space within which the balloon is depicted as expanding, does contain a point in its interior, which is central to the balloon's surface, but this point is not itself *part* of the balloon's surface, where that surface is taken to represent the entirety of the universe's spatial geometry.

The time-dependence of the *actual* universe's expansion, that observations reveal, is indeed in striking accordance with the equations of Einstein's general theory of relativity, but apparently only if two somewhat unexpected ingredients are incorporated into the theory, now commonly referred to under the (somewhat unfortunate[2.4]) names of 'dark matter' and 'dark energy'. Both of these ingredients will have considerable importance for the proposed scheme of things that I shall be introducing the reader to in due course (see §3.1, §3.2). They are now part of the standard picture of modern cosmology, though it must be said that neither is completely accepted by all experts in the field.[2.5] For my own part, I am happy to accept *both* the presence of some invisible material—the 'dark matter'—of a nature that is essentially unknown to us, yet constituting some 70% of the material substance of our universe, and also that Einstein's equations of general relativity must be taken in the *modified* form that he himself put forward in 1917 (though he later retracted it), in which a tiny positive *cosmological constant* Λ (the most plausible form of 'dark energy') must be incorporated.

It should be pointed out that Einstein's general theory of relativity (with or without the tiny Λ) is now extremely well tested at the scale of the solar system. Even the very practical global positioning devices, that are now in common use, depend upon general relativity for their remarkable accuracy of operation. Considerably more impressive is the extraordinary precision of Einstein's theory in its modelling of the behaviour of binary pulsar[2.6] systems—up to an overall precision of something like one part in 10^{14} (in the sense that the timing of the pulsar signals from the binary system PSR-1913+16, over a period of some 40 years, is accurately modelled with a precision of around 10^{-6} of a second per year).

The original cosmological models, based on Einstein's theory, were those put forward by the Russian mathematician Alexander Friedmann in 1922 and 1924. In Fig. 2.2, I have sketched the space-time histories of these models, depicting the time evolutions of the three cases (taking $\Lambda=0$), in which the *spatial* curvature of the universe is, respectively, positive, zero, and negative.[2.7] As will be my convention, in virtually all my space-time diagrams, the vertical direction represents *time* evolution and the horizontal directions, space. In all three cases, it is assumed that the spatial part of the geometry is completely uniform (what is called homogeneous and isotropic). Cosmological models with this kind of symmetry are called *Friedmann–Lemaître–Robertson–Walker* (FLRW) models. The original Friedmann models are a particular case, where the type of matter being described is a *pressureless fluid*, or 'dust' (see also §2.4).

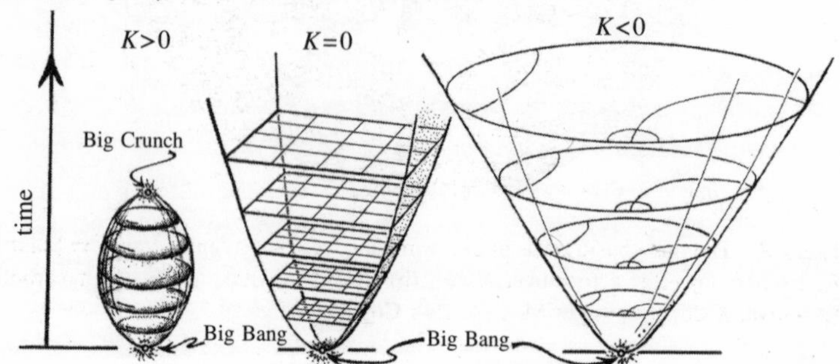

Fig. 2.2 Space-time histories for Friedmann's cosmological models in which the spatial curvature of the universe is positive, zero, and negative (left to right).

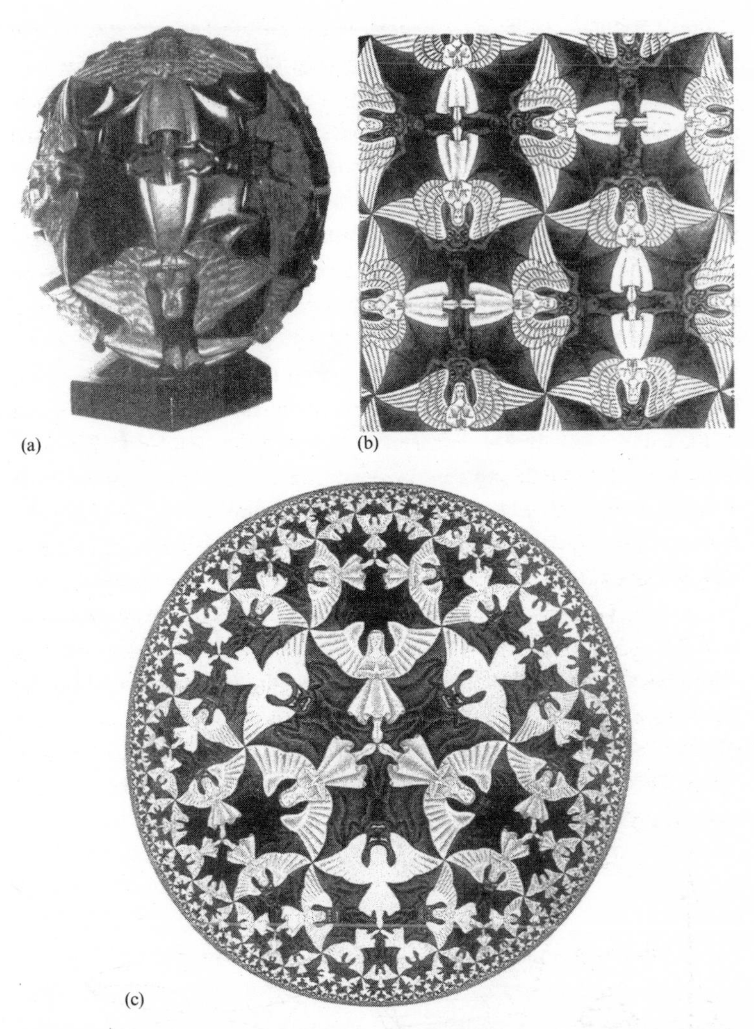

Fig. 2.3 The three basic kinds of uniform plane geometry as illustrated by Maurits C. Escher: (a) elliptic (positive, $K>0$); (b) Euclidean (flat, $K=0$); (c) hyperbolic (negative, $K<0$). Copyright M. C. Escher Company (2004).

Essentially,[2.8] there are just these three cases to consider for the spatial geometry, namely the case $K>0$ of positive spatial curvature, where the spatial geometry is the 3-dimensional analogue of a spherical surface (like our balloon, referred to above), the flat case $K=0$, where the spatial geometry is the familiar 3-dimensional geometry of Euclid, and the negatively curved case $K<0$ of *hyperbolic* spatial 3-geometry. It is fortunate for us that the Dutch artist Maurits C. Escher has illustrated all three of these different kinds of geometry beautifully in terms of tessellations of angels and devils, see Fig. 2.3. We must bear in mind that these simply depict *2-dimensional* spatial geometry, but analogues of all three kinds of geometry exist also in the full 3 spatial dimensions.

All these models originate with a 'Big-Bang' singular state—where 'singular' refers to the fact that the density of material and the curvature of the space-time geometry become infinite at this initial state—so that Einstein's equations (and physics, as a whole, as we know it) simply 'give up' at the singularity (although see §3.2 and Appendix B10). It will be noted that the temporal behaviour of these models rather mirrors their spatial behaviour. The spatially finite case ($K>0$; Fig. 2.3(a)) is also the temporally finite case, where not only is there an initial Big-Bang singularity but there is also a *final* one, commonly referred to as a 'Big Crunch'. The other two cases ($K\leq 0$; Fig. 2.3(b),(c)) are not only spatially infinite[2.9] but temporally infinite also, their expansion continuing indefinitely.

Since around 1998, however, when two observational groups, one headed by Saul Perlmutter and the other by Brian P. Schmidt, had been analysing their data concerning very distant supernova explosions,[2.10] evidence has mounted which strongly indicates that the expansion of the universe in its later stages does not actually match the evolution rates predicted from the standard Friedman cosmologies that are illustrated in Fig. 2.2. Instead, it appears that our universe has begun to accelerate in its expansion, at a rate that would be explained if we are to include into Einstein's equations a cosmological constant Λ, with a small positive value. These, and later observations of various kinds,[2.11] have provided fairly convincing evidence of the beginnings of the *exponential expansion* characteristic of a Friedmann model with $\Lambda > 0$. This exponential expansion occurs not only with the cases $K\leq 0$ which in any case expand indefinitely in their remote futures even when $\Lambda = 0$, but also in the spatially closed case $K>0$, provided that Λ is large enough to overcome the tendency for recollapse that the closed Friedmann model possesses. Indeed, the evidence does indicate the presence of a large-enough Λ, so that the value (sign) of K has become more-or-less irrelevant to the expansion rate, where the (positive) value of Λ that appears actually to be present in Einstein's equations would then dominate the late-time behaviour, providing an exponential expansion independently of the value of K within the observationally acceptable range. Accordingly, we appear to have a universe with an expansion rate that is basically in accordance with the curve shown in Fig. 2.4, the space-time picture appearing to be in accordance with Fig. 2.5.

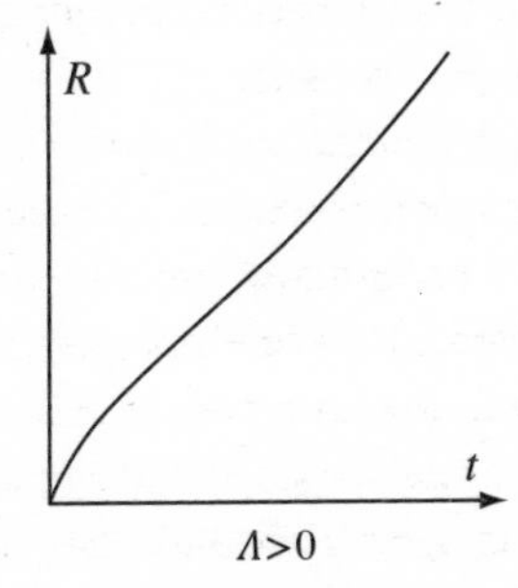

Fig. 2.4 Expansion rate of the universe for positive Λ, with eventual exponential growth.

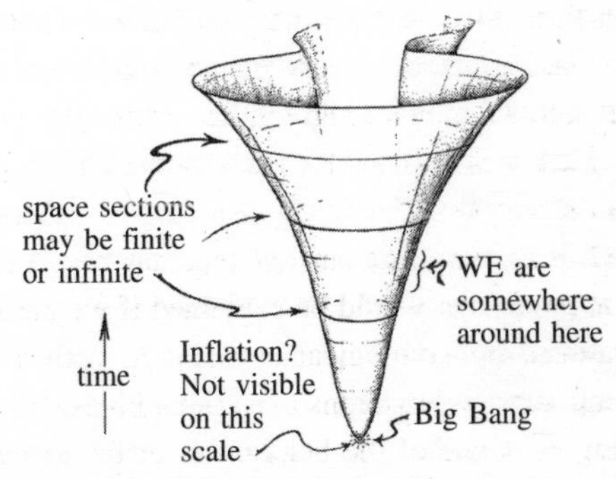

Fig. 2.5 Space-time expansion of the universe. Picture with positive Λ (suggestively drawn so as not be be biased as to the value of K).

In view of this, I shall not be particularly concerned here with the difference between these three possibilities for the universe's spatial geometry. In fact the present observations indicate an overall spatial geometry for the universe that is rather close to the flat case $K=0$. In one sense this is somewhat unfortunate, because it tells us that we really do not know the answer to the question of what the overall spatial geometry of the universe actually is likely to be—whether the universe is necessarily spatially closed, or might be spatially infinite, for example—because in the absence of powerful theoretical reasons for believing the contrary, there will always remain some possibility of a small positive or negative overall curvature.

On the other hand, many cosmologists are of the opinion that the viewpoint provided by *cosmic inflation* does provide a powerful reason for believing that the spatial geometry of the spatial universe must be (apart from relatively small local deviations) actually *flat* ($K=0$), so they are pleased by this observational closeness to flatness. Cosmic inflation is a proposal that, within a very tiny time-period somewhere between around 10^{-36} and 10^{-32} seconds after the Big Bang, the universe underwent an exponential expansion, increasing its linear dimension by an enormous factor of around 10^{30} or 10^{60} (or even 10^{100}) or so. I shall have more to say about cosmic inflation later (see §2.6), but for the moment, I should just warn the reader that I am not enthusiastic about this particular proposal, despite its largely universal acceptance among present-day cosmologists. In any case, the presence of an early inflationary stage in the history of the universe would not affect the appearance of Figs. 2.2 and 2.5, since the effects of inflation would show up only at the very early stages, just following the Big Bang, and would not be visible on the scale at which Figs. 2.2 and 2.5 are drawn. On the other hand, the ideas that I shall be putting forward later in this book appear to provide credible *alternatives* to inflation for explaining those observed phenomena that *seem* to depend upon it in the currently popular cosmological schemes (see §3.5).

Apart from such considerations, I have a quite different motivation for presenting the picture of Fig. 2.3(c) here, since it illustrates a point which will have fundamental significance for us later on. This beautiful Escher print is based on a particular representation of the hyperbolic plane which is one of several put forward by the highly ingenious Italian geometer Eugenio Beltrami[2.12] in 1868. The same representation was rediscovered, about 14 years later, by the leading French mathematician Henri Poincaré, whose name is more commonly attached to it. To avoid adding to this confusion about terminology, I shall usually refer to it here simply as the *conformal* representation of the hyperbolic plane, the term 'conformal' referring to the fact that *angles* in this geometry are correctly represented in the Euclidean plane in which it has been depicted. The ideas of conformal geometry will be addressed in a little more detail in §2.3.

We are to think of all the devils in the geometry as being *congruent* with each other according to the hyperbolic geometry being represented, and likewise all the angels to be regarded as congruent. Clearly their *sizes*, according to the background Euclidean measure, are represented as tinier the closer to the circular boundary we examine them, but the representation of *angles* or infinitesimal *shapes* remain true, as close to the boundary as we care to examine them. The circular boundary itself represents *infinity* for this geometry, and it is this *conformal representation of infinity* as a smooth finite *boundary* that I am pointing out here to the reader, as it will be playing a central role in the ideas that we shall be coming to later (particularly in §2.5 and §3.2).

2.2 The ubiquitous microwave background

In the 1950s, a popular theory of the universe was one referred to as the *steady state* model, a proposal first put forward by Thomas Gold and Hermann Bondi in 1948, and soon taken up in more detail by Fred Hoyle,[2.13] who were all at Cambridge University at the time. The theory required there to be a continual creation of material throughout space, at an extremely low rate. This material would have to be in the form of hydrogen molecules—each being a pair consisting of one proton and one electron, created out of the vacuum—at the extremely tiny rate of about one such atom per cubic metre per thousand million years. This would have to be at just the right rate to replenish the reduction of the density of material due to the expansion of the universe.

In many respects, this is a philosophically attractive and aesthetically pleasing model, as the universe requires no origin in time or space, and many of its properties can be deduced from the requirement that it should be self-propagating. It was fairly soon after this theory was being proposed that I entered Cambridge University, in 1952, as a young graduate student (researching in pure mathematics, but with a keen interest in physics and cosmology[2.14]), and I returned later, in 1956, as a research fellow. While at Cambridge, I got to know all three of the steady-state theory's originators, and I had certainly found this model to be appealing and the arguments fairly persuasive. However, towards the end of my time at Cambridge, detailed counts of distant galaxies carried out at the Mullard Radio

Observatory by (Sir) Martin Ryle (also in Cambridge) were beginning to provide clear observational evidence *against* the steady-state model.[2.15]

But the real death-blow was the accidental observation by the Americans Arno Penzias and Robert W. Wilson, in 1964, of microwave electromagnetic radiation, coming from all directions in space. Such radiation had in fact been predicted, in the later 1940s by George Gamow, and by Robert Dicke on the basis of what was then the more conventional 'Big-Bang theory', such presently observable radiation being sometimes described as 'the flash of the Big Bang', the radiation having been cooled from some 4000 K to a few degrees above absolute zero[2.16] by an enormous red-shift effect due to the vast expansion of the universe since the emission of the radiation. After Penzias and Wilson had convinced themselves that the radiation they were observing (of around 2.725 K) was genuine, and must actually be coming from deep space, they consulted Dicke, who was quick to point out that their puzzling observations could be explained as what he and Gamow had previously predicted. This radiation has gone under various different names ('relic radiation', 3-degree background, etc.); nowadays it is commonly referred to simply as the 'CMB', which stands for 'cosmic microwave background'.[2.17] In 1978, Penzias and Wilson were awarded the Nobel Prize in Physics for its discovery.

The source of the photons which actually constitute the CMB that we now 'see' is not really the 'actual Big Bang', however, as these photons come to us directly from what is called the 'surface of last scattering' which occurred some 379 000 years following the moment of the Big Bang (i.e. when the universe was about 1/36 000 of its present age). Earlier than this, the universe was opaque to electromagnetic radiation because it would have been inhabited by large numbers of separate charged particles—mainly protons and electrons—milling around separately from each other, constituting what is referred to as a 'plasma'. Photons would have scattered many times in this material, being absorbed and created copiously, and the universe would have been very far from transparent. This 'foggy' situation would have continued until the time referred to as 'decoupling' (where 'last scattering' occurs) at which the universe became transparent because it had cooled down sufficiently for the separate electrons and protons to be able to pair up, largely in the form of hydrogen (with a few other atoms produced, mainly about 23% helium, whose nuclei—called 'α-particles'—would have been among the products of the first few minutes of the universe's existence). The photons were then able to decouple from these neutral atoms, to travel essentially undisturbed from then on, to become the radiation which is now perceived as the CMB.

Since its initial observation in the 1960s, many experiments have been performed to get better and better data concerning the nature and distribution of the CMB, there being so much detailed information now, that the subject of cosmology has been completely transformed—from one in which there was much speculation and very little data to bear on this speculation—to a *precision science*, in which, although there is still much speculation, there is now a very great deal of detailed data to modulate this speculation! One particularly noteworthy experiment was the COBE satellite (Cosmic Background Explorer), launched by NASA in November 1989. Its remarkable observations earned George Smoot and John Mather the 2006 Nobel Prize in Physics.

There are two very striking and important features of the CMB which were made particularly evident by COBE, and I want to concentrate on both of these. The first is the extraordinary closeness by which the observed frequency spectrum matches that explained by Max Planck in 1900 to account for the nature of what is called 'black-body radiation' (and which marked the starting point of quantum mechanics). The second is the extremely uniform nature of the CMB over the whole sky. Each of these two facts will be telling us something very fundamental about the nature of the Big Bang, and its curious relation to the Second Law. Much of modern cosmology has moved on from this now, and is concerned more with the slight and subtle *deviations* from uniformity in the CMB that are also seen. I shall be coming to some of these later (see §3.6), but for the moment I shall need to address these two more blatant facts in turn, as we shall find that they both have a very great significance for us.

Figure 2.6 depicts the frequency spectrum of the CMB, essentially as initially measured by COBE, but where now greater precision is obtained from later observations. The vertical axis measures the *intensity* of the radiation, as a function of the different frequencies, these being marked off along the horizontal axis with increasing frequency off to the right.

The continuous line is Planck's 'black-body curve', which is given by a specific formula,[2.18] and it is what quantum mechanics tells us is the radiation spectrum of *thermal equilibrium*, for any particular temperature T. The little vertical bars are *error bars*, telling us roughly the range within which the observed intensities lie. It should be noted, however, that these error bars are exaggerated by a factor of 500, so the *actual* observation points lie much more closely on the Planck curve than would appear—in fact, so closely that, to the eye, even the observations on the very far right, where the error is greatest, concur with the Planck curve to within the thickness of the ink line! Indeed, the CMB provides us with the most precise agreement between an observed intensity spectrum and the calculated Planck black-body curve that is known in observational science.

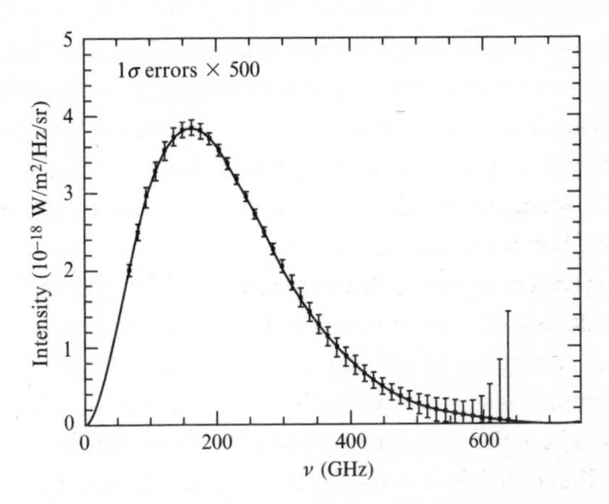

Fig. 2.6 Frequency spectrum of the CMB, as initially observed by COBE, but supplemented by later more precise observations. Note that the 'error bars' are exaggerated by a factor of 500. This shows precise agreement with the Planck spectrum.

What does this tell us? It appears to tell us that what we are looking at comes from a state that must effectively be thermal equilibrium. But what does 'thermal equilibrium' actually mean? I refer the reader back to Fig. 1.15, where we find the words 'thermal equilibrium' labelling the coarse-graining region of phase space which is (by far) the largest of all. In other words, this is the region representing *maximum* entropy. But we must recall the thrust of the arguments given in §1.6. These arguments told us that the whole basis of the Second Law must be explained by the fact that the initial state of the universe—which we evidently must take as being the Big Bang—must be a (macroscopic) state of extraordinarily *tiny* entropy. What we appear to have found is essentially the complete opposite, namely a (macroscopic) state of *maximum* entropy!

One point must be addressed here, namely the fact that the universe is *expanding*, so what we are looking at can hardly be an actual 'equilibrium' state. However, what is evidently happening here is an *adiabatic* expansion, where 'adiabatic' here refers, effectively, to a 'reversible' change in which the entropy remains constant. The fact that this kind of 'thermal state' is actually preserved in the early universe's expansion was pointed out by R.C. Tolman in 1934.[2.19] We shall be seeing some more of Tolman's contributions to cosmology in §3.3. In terms of phase space, the picture is more like Fig. 2.7 than Fig. 1.15, where the expansion is described as a succession of maximal coarse-graining regions of essentially equal volume. In this sense, the expansion can still be viewed as a kind of thermal equilibrium.

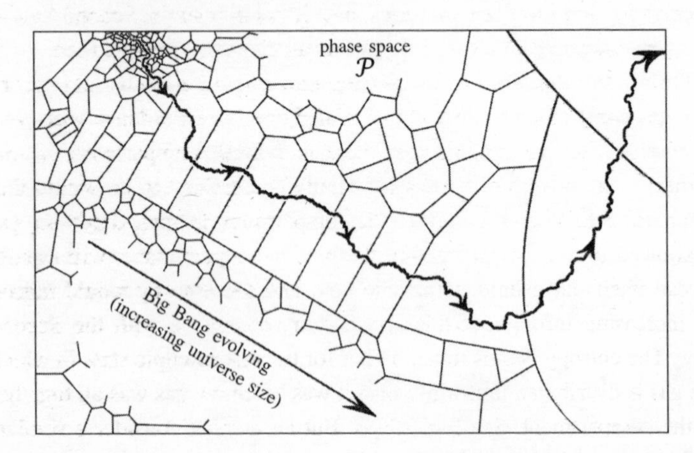

Fig. 2.7 Adiabatic expansion of the universe depicted as a succession of maximal coarse-graining regions of equal volume.

So we still seem to be seeing *maximum* entropy. Something appears to have gone seriously wrong with the arguments. It is not even just that the observations of the universe have come up with a surprise. Not at all: in a certain sense, the observations are closely in accord with what was expected. Given that there *was* actually a Big Bang, and that this initial state is to be described as being in accord with the standard picture presented by general-relativistic cosmology, then a very hot and uniform initial thermal state *is* what would be expected. So where does the resolution of this conundrum lie? Perhaps rather surprisingly, the issue has to do with the *assumption* that the universe is indeed in accordance with the standard picture of relativistic cosmology! We shall need to examine this assumption very carefully indeed, to see what has eluded us.

First, we must remind ourselves what Einstein's general theory of relativity is all about. It is, after all, an extraordinarily accurate theory of *gravity*, where the gravitational field is described in terms of a curvature of space-time. I shall have a lot to say about this theory in due course, but for the moment let us think in terms of the older—and still extraordinarily accurate—*Newtonian* gravitational theory, and try to understand, in rough general terms, how it fits in with the Second Law—of *thermodynamics*, that is; I do not mean *Newton's* second law!

Often, considerations of the Second Law might be discussed in terms of a gas constrained to lie within a sealed box. In accordance with such discussions, let us imagine that there is a small compartment in one corner of the box, and the gas is initially constrained to be within that compartment. When the door to the compartment is opened and the gas is allowed to move freely within the box, we expect that it will rapidly spread itself out evenly within the box, and the entropy would indeed be increasing throughout this process, in accordance with the Second Law. The entropy is thus much higher for the macroscopic state in which the gas is distributed uniformly than it was when the gas was all together in the compartment. See Fig. 2.8(a). But let us now consider a similar-looking situation, but with an imaginary box of galactic size, and where the individual molecules of gas are replaced by individual stars moving within this box. The difference between this situation and that of the gas is not just a matter of scale, and I shall take size to be irrelevant for the present purposes. What is relevant is the fact that the stars *attract* each other, through the relentless force of gravity. We might imagine that the distribution of stars is *initially* spread fairly uniformly throughout our galactic-sized box. But, now, as time progresses, we find a tendency for the stars to collect together in clumps (and generally to move more rapidly as they do so). Now the uniform distribution is *not* the one of highest entropy, increasing entropy being accompanied by an increase in the clumpiness of the distribution. See Fig. 2.8(b).

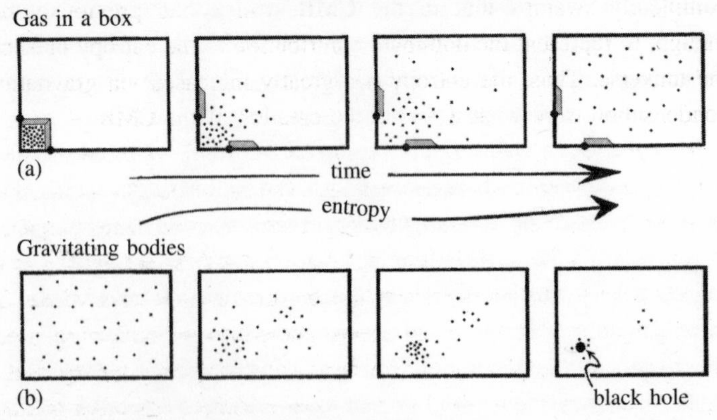

Fig. 2.8 (a) Gas is initially constrained within a small compartment in the corner of a box before being released and distributing itself uniformly throughout the box. (b) In a galactic-sized box, stars are initially uniformly distributed but collect together in clumps over time: a uniform distribution in this case is not the one with highest entropy.

We may ask what now is the analogue of thermal equilibrium, where the entropy has increased to its maximum? It turns out that this question cannot be properly addressed within the confines of Newtonian theory. If we consider a system that consists of massive point particles attracting each other according to Newton's inverse square law, then we can envisage states in which some of the particles get progressively closer and closer to each other, moving more and more rapidly, so that there is no limit to the degree of clumpiness and rapidity of motion, and the proposed state of 'thermal equilibrium' simply does not exist. The situation turns out to be much more satisfactory in Einstein's theory, because the 'clumpiness' can *saturate*, when the matter conglomerates into a *black hole*.

We shall come to black holes in more detail in §2.4, where we learn that the formation of a black hole represents an enormous increase in the entropy. Indeed, at the present epoch of the universe's evolution, the greatest entropy contribution, by far, lies in large black holes, like the one at the centre of our own Milky Way galaxy, with a mass of around 4 000 000 times the mass of our Sun. The total entropy in such objects completely swamps that in the CMB, which had previously been thought to represent the dominant contribution to the entropy present in the universe. Thus, the entropy has greatly increased via gravitational condensation from what it was at the creation of the CMB.

This relates to the *second* feature of the CMB referred to above, namely its closely uniform temperature over the whole sky. How closely uniform is it? There is a slight temperature variation understood as a Doppler shift, coming from the fact that the Earth is not exactly at rest with respect to the mass distribution of the universe as a whole. The Earth's motion is composed of various contributions, such as its motion about the Sun, the Sun's motion around the Milky-Way galaxy, and the galaxy's motion due to local gravitational influences of other relatively nearby mass distributions. All combine together to provide what is referred to as the Earth's 'proper motion'. This leads to a very slight increase of the apparent temperature of the CMB in the direction in the sky that we are moving towards,[2.20] and a very slight decrease in the direction in the sky that we are moving away from, and an easily calculated pattern of slight temperature alterations over the whole sky. Correcting for this, we find a CMB sky that has an extraordinarily uniform temperature over the sky, with deviations of the order of only a few parts in 10^5.

This tells us that, at least over the surface of last scattering, the universe was extraordinarily uniform, like the *right*-hand picture of Fig. 2.8(a) and also like the *left*-hand picture Fig. 2.8(b). It is reasonable to assume, therefore, that so long as we can ignore the influences of gravity, the *material* content of the universe (at last scattering) was indeed at as high an entropy as it could achieve on its own. Gravitational influences would, after all, be small *because* of the uniformity, but it was this very uniformity in the matter distribution that provided the *potential* for enormous subsequent entropy increases when gravitational influences later come into play. Our picture of the entropy of the Big Bang is therefore completely changed once we consider the introduction of gravitational degrees of freedom. It is the *assumption* that our universe, overall, is very closely in accord with the spatial homogeneity and isotropy—sometimes referred to as the 'cosmological principle'[2.21], basic to FLRW cosmology and, in particular, central to the Friedmann models discussed in §2.1—that implies the huge suppression of gravitational degrees of freedom in the initial state. This early spatial uniformity represents the universe's extraordinarily low initial entropy.

A natural question to ask is: what on earth does that cosmological uniformity have to do with our familiar Second Law, which seems to permeate so much of the detailed physical behaviour in the world we know? There are multitudes of commonplace instances of the Second Law that would seem to bear no relation to the fact that gravitational degrees were suppressed in the early universe. Yet the connection is indeed there, and it is actually not so hard to trace back these commonplace instances of the Second Law to the uniformity of the early universe.

Let us consider, as an example, the egg of §1.1, perched on the edge of a table, about to fall off and smash on the floor below (see Fig. 1.1). The entropy-raising process of the egg rolling off the table and smashing is enormously favoured, probabilistically, provided that we are prepared to *assume* that the egg started in the very low-entropy state of being perched, unbroken, at the edge of the table. The puzzle of the Second Law is not the raising of the entropy following that event; the puzzle lies in the event itself, i.e. the question of how the egg happened to find itself in this extremely low-entropy state in the first place. The Second Law tells us that it must have arrived in this very improbable state through a sequence of other states that had been even more improbable prior to this, and getting more so the farther back in time we examine the system.

There are basically two things to explain. One is the question of how the egg got up on the table, and the other is how the low-entropy structure of that egg itself came about. Indeed, the material of an egg (taken to be a hen's egg) has been superbly organized into a perfect package of appropriate nourishment for an intended chick. But let us start with what may seem to be the easier part of the problem, namely that of how the egg found itself up on the table. The likely answer would be that some person put it there, perhaps a little carelessly, but human intervention was the probable cause. There is clearly a lot of highly organized structure in a functioning human being, which suggests a low entropy, and the placing of the egg on the table would have taken only a very little from the rather large reservoir of low entropy in the relevant system, consisting of a reasonably well-fed person and a surrounding oxygen-laden atmosphere. The situation with the egg itself is somewhat similar, in that the egg's highly organized structure, superbly geared to supporting the burgeoning life of a presumed embryo within it, is very much part of the grand scheme of things that keeps life going, on this planet. The entire fabric of life on Earth requires the maintaining of a profound and subtle organization, which undoubtedly involves entropy being kept at a low level. In detail, there is an immensely intricate and interconnected structure, which has evolved in keeping with the fundamental biological principle of natural selection and with many detailed matters of chemistry.

What, you might well ask, do such matters of biology and chemistry have to do with the uniformity of the early universe? Biological complication does not allow the system as a whole to violate the general laws of physics, such as the law of conservation of energy; moreover, it cannot provide escape from the constraints imposed by the Second Law. The structure of life on this planet would run rapidly down were it not for a powerful low-entropy source, upon which almost all life on Earth depends, namely the *Sun*.[2.22] One tends to think of the Sun as supplying the Earth with an external source of *energy*, but this is not altogether correct, as the energy that the Earth receives from the Sun by day is essentially *equal* to that which the Earth returns to the darkness of space![2.23] If this were not so, then the Earth would simply heat up until it reaches such an equilibrium. What life depends upon is the fact that the Sun is much hotter than the darkness of space, and consequently the photons from the Sun have a considerably higher frequency (namely that of yellow light) than the infra-red photons that Earth returns to space. Planck's formula $E=h\nu$

(see §2.3) then tells us that, on average, the energy carried in by individual photons from the Sun is considerably greater than the energy carried out by individual photons returning to space. Thus, there are many more photons carrying the same energy away from Earth than there are that carry it in from the Sun. See Fig. 2.9. More photons imply more degrees of freedom and therefore a larger phase-space volume. Accordingly, Boltzmann's $S = k \log V$, (see §1.3) tells us that energy coming in from the Sun carries a considerably lower entropy than that returning to space.

Fig. 2.9 Photons arriving at the Earth's surface from the Sun have higher energy (shorter wavelength) than those returned to space by the Earth. Given an overall energy balance (the Earth does not get hotter over time), there must be more photons leaving than arriving; that is, the energy arriving has lower entropy than that departing.

Now, on Earth, the green plants have, by the process of photosynthesis, found a way of converting the relatively high-frequency photons coming from the Sun to photons of a lower frequency, using this gain in low entropy to build up their substance by extracting carbon from CO_2 in the air and returning it as O_2. When animals eat plants (or eat other animals that eat plants), they use this source of low entropy, and the O_2, to keep down their own entropy.[2.24] This applies to humans, of course, and also to chickens, and it supplies the source of low entropy needed for the construction of our unbroken egg and for it to be placed on the table!

So what the Sun does for us is not simply to supply us with energy, but to provide this energy in a low-entropy form, so that we (via the green plants) can keep our entropy down, this coming about because the Sun is a *hot spot in an otherwise dark sky*. Had the entire sky been of the same temperature as that of the Sun, then its energy would have been of no use whatever to life on Earth. This applies, also, to the Sun's ability to raise water from the oceans high up into the clouds, which again depends crucially on this temperature difference.

Why is the Sun a hot spot in the dark sky? Well, there are all sorts of complicated processes going on in the Sun's interior, and the thermonuclear reactions that result in hydrogen being converted to helium play an important part in this. However the key issue is that the Sun is there at all, and this has come about from the gravitational influence which holds the Sun together. Without thermonuclear reactions, the Sun would still shine, but shrink and get much hotter, and have a far shorter life. On Earth, we clearly gain from these thermonuclear reactions, but they would not even have the chance to take place were it not for the gravitational clumping that produced the Sun in the first place. Accordingly, it is the potential for stars to form (albeit via somewhat complicated processes in appropriate regions in space), through the relentless entropy-raising process of gravitational clumping, from initial material that started off in a very uniform gravitationally low-entropy state.

This all comes about, ultimately, from the fact that we have been presented with a Big Bang of a very special nature, the extreme (relative) *lowness* of its entropy being manifested in the fact that its gravitational degrees of freedom were indeed not initially activated. This is a curiously lop-sided situation, and to understand it better we shall try to dig a little more deeply, in the next three sections, into Einstein's beautiful curved-space-time description of gravity. Then, in §2.6 and §3.1, I shall return to the issue of the nature of this extraordinary specialness that is actually exhibited in our Big Bang.

2.3 Space-time, null cones, metrics, conformal geometry

When, in 1908, the distinguished mathematician Hermann Minkowski—who had coincidentally been one of Einstein's teachers at the Zurich Polytechnic—demonstrated that he could encapsulate the basics of special relativity in terms of an unusual type of 4-dimensional geometry, Einstein was less than enthusiastic about the idea. But later he realized the crucial importance of Minkowski's geometric notion of *space-time*. Indeed, it formed an essential ingredient of his own generalization of Minkowski's proposal to provide the curved space-time basis of his *general* theory of relativity.

Minkowski's 4-space incorporated the standard three dimensions of space with a fourth dimension to describe the passage of time. Accordingly, the *points* of this 4-space are frequently referred to as *events*, since any such point has a temporal as well as a spatial specification. There is not really anything very revolutionary about this, just in itself. But the key point of Minkowski's idea—which *was* revolutionary—is that the geometry of his 4-space does not separate out naturally into a time dimension and (more significantly) a family of ordinary Euclidean 3-spaces, one for each given time. Instead, Minkowski's space-time has a different kind of geometric structure, giving a curious twist to Euclid's ancient idea of geometry. It provides an *overall* geometry to space-time, making it one indivisible whole, which completely encodes the structure of Einstein's special relativity.

Thus, in Minkowski's 4-geometry, we are *not* now to think of the space-time as being simply built out of a succession of 3-surfaces, each representing what we think of as 'space' at various different times (Fig. 2.10). For that interpretation, each of these 3-surfaces would describe a family of events all of which would be taken to be *simultaneous* with one another. In special relativity, the notion of 'simultaneous' for spatially separated events does not have an absolute meaning. Instead, 'simultaneity' would depend upon some arbitrarily chosen observer's velocity.

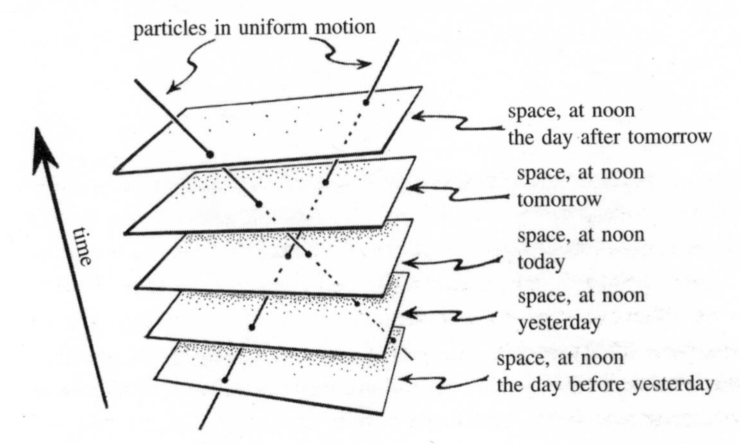

Fig. 2.10 Space-time before Minkowski.

This, of course, is at odds with common experience, for we *do* seem to have a notion of simultaneity for distant events that is independent of our velocity. But (according to Einstein's special relativity) if we were to move at a speed that is comparable with that of light, then events that seem to us to be simultaneous would generally not seem to be simultaneous to some other such observer, with a different velocity. Moreover, the velocities would not even have to be very large if we are concerned with very *distant* events. For example, if two people stroll past each other in opposite directions along a path, then the events on the Andromeda Galaxy that they would each individually consider to be simultaneous with that particular event at which they pass one another would be likely to differ by several weeks,[2.25] see Fig. 2.11!

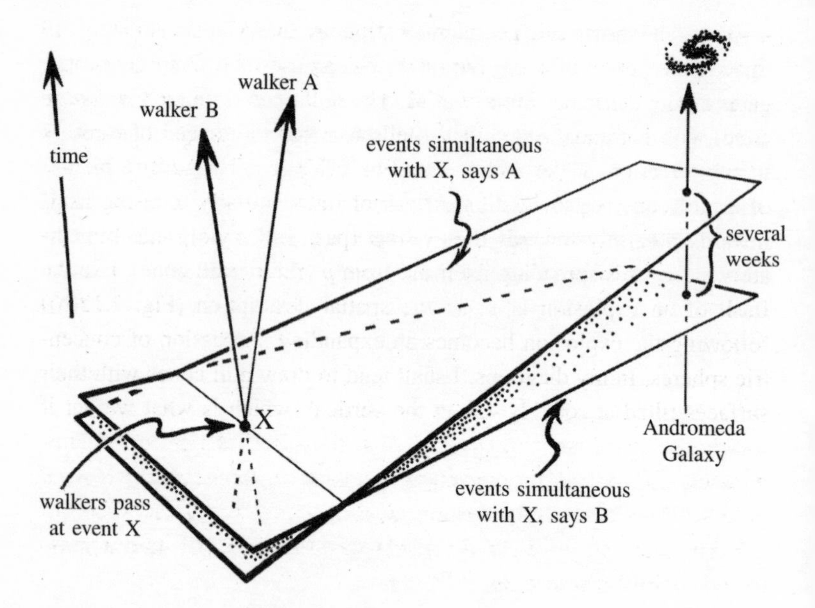

Fig. 2.11 Two walkers amble past one another, but the event X of their passing is judged by each to be simultaneous with events on Andromeda differing by several weeks.

According to relativity, the notion of 'simultaneous', for distant events, is not an absolute thing, but depends upon some observer's velocity to be specified, so the slicing of space-time into a family of simultaneous 3-spaces is *subjective* in the sense that for a different observer velocity we get a different slicing. What Minkowski's space-time achieves is to provide an *objective* geometry, that is not dependent on some arbitrary observer's view of the world, and which does not have to change when one observer is replaced by another. In a certain sense, what Minkowski did was to take the 'relativity' out of special relativity theory, and to present us with an *absolute* picture of spatio-temporal activity.

But for this to give us a firm picture, we need a kind of *structure* for the 4-space to replace the idea of a temporal succession of 3-spaces. What structure is this? I shall use the letter $\mathbb{M}$ to denote Minkowski's

4-space. The most basic geometrical structure that Minkowski assigned to 𝕄 is the notion of a *null cone*,[2.26] which describes how light propagates at any particular event p in 𝕄. The null cone—which is a *double* cone, with common vertex at p—tells us what the 'speed of light' is in any direction, at the event p (see Fig. 2.12(a)). The intuitive picture of a null cone is provided by a flash of light, initially focusing itself inwards precisely towards the event p (past null cone), and immediately afterwards spreading itself out from p (future null cone), like the flash of an explosion at p, so the spatial description (Fig. 2.12(b)) following the explosion becomes an expanding succession of concentric spheres. In my diagrams, I shall tend to draw null cones with their surfaces tilted at roughly 45° to the vertical, which is what we get if we choose space and time *units* so that the speed of light $c=1$. Thus if we choose *seconds* for our time scale, then we choose a *light-second* (=299 792 458 metres) for our unit of distance; if we choose years for our time scale, then we choose a light-year ($\cong 9.46 \times 10^{12}$ kilometres) for our unit of distance, etc.[2.27]

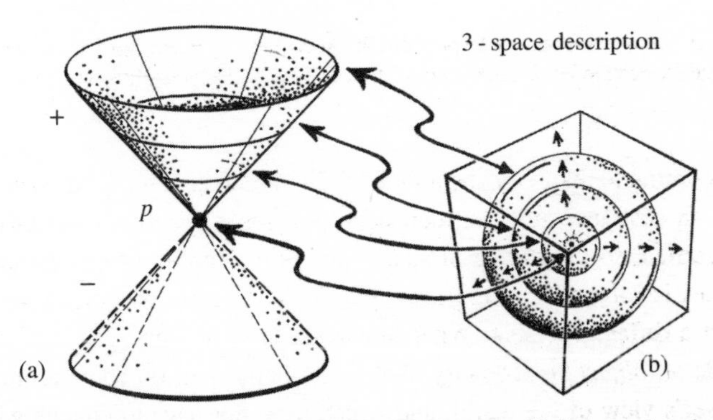

Fig. 2.12 (a) Null cone at p in Minkowski's 4-space; (b) 3-space description of the future cone as an expanding succession of concentric spheres originating at p.

Einstein's theory tells us that the speed of any massive particle must always be less than that of light. In space-time terms, this means that the *world-line* of such a particle—this being the locus of all the events that constitute the particle's history—must be directed *within* the null cone at each of its events. See Fig. 2.13. A particle may have a motion that is *accelerated* at some places along its world-line, whence its world-line need not be straight, the acceleration being expressed, in space-time terms, as a *curvature* of the world-line. Where the world-line is curved, it is the *tangent vector* to the world-line that must lie within the null cone. If the particle is mass*less*,[2.28] such as a photon, then its world-line must lie *along* the null cone at each of its events, since its speed at every one of its events is indeed taken to be light speed.

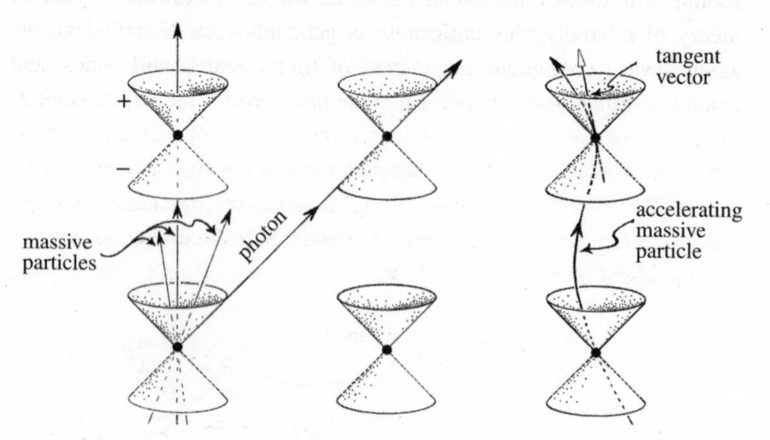

Fig. 2.13 Null cones in 𝕄, uniformly arranged. World lines of massive particles are directed within the cones and of massless ones along the cones.

The null cones also tell us about *causality*, which is the issue of determining which events are to be regarded as being able to influence which other events. One of the tenets of (special) relativity theory is the assertion that signals shall not be allowed to propagate faster than light. Accordingly, in terms of the geometry of 𝕄, we say that an event p would be permitted to have a causal influence on event q if there is a world-line connecting p to q, that is a (smooth) path from p to q lying on or within the null cones. For this, we need to specify an *orientation* to the path (indicated by attaching an 'arrow' to the path), that proceeds uniformly from past to future. This requires that 𝕄's geometry be assigned a *time orientation*, which amounts to a consistent continuous separate assignment of 'past' and 'future' to the two components of each null cone. I have labelled a past component with a '−' sign and a future component with a '+' sign. This is illustrated in Figs. 2.12(a) and 2.13, where the past null cone is distinguished in my drawings by the use of broken lines. The normal terminology of 'causation' takes causal influences to proceed in the past-to-future direction, i.e. along world-lines whose oriented tangent vectors point on or within *future* null cones.[2.29]

𝕄's geometry is completely uniform, where each event is on an equal footing with every other event. But when we pass to Einstein's *general* theory of relativity, this uniformity is generally lost. Nonetheless, we again have a continuous assignment of time-oriented null cones, and again it is true that any massive particle has a world-line whose (future-oriented) tangent vectors all lie within these future null cones. And, as before, a massless particle (photon) has a world-line whose tangent vectors all lie along null cones. In Fig. 2.14 I have depicted the kind of situation which occurs in general relativity, where the null cones are not now arranged in a uniform fashion.

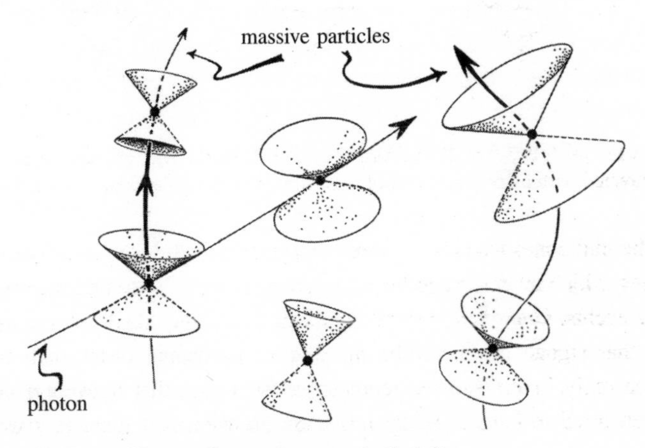

Fig. 2.14 Non-uniform null cones in general relativity.

We have to try to think of these cones being drawn on some kind of ideal 'rubber sheet' with the null cones printed on it. We can move the rubber sheet around and distort it in any way we like, so long as the deformation is done in a smooth way, where the null cones are carried around with the rubber sheet. Our null cones determine the 'causality structure' between events, and this is not altered by any such deformation provided that the cones are thought of as being carried around with the sheet.

A somewhat analogous situation is provided by Escher's depiction of the hyperbolic plane shown in Fig. 2.3(c), in §2.1, where we can imagine Escher's picture to be printed on such an ideal rubber sheet. We might choose one of the devils that appears to be close to the boundary, and move it, by such a smooth deformation of the sheet, so that it comes into the location previously occupied by one near the centre. This motion can be made to carry all the devils into locations previously occupied by other devils, and such a motion would describe a symmetry of the underlying hyperbolic geometry illustrated by Escher's picture. In general relativity, symmetries of this kind can occur (as with the Friedmann models described in §2.1), but this is rather exceptional. However, the possibility of carrying out such 'rubber-sheet' deformations is very much part of the general theory, these being referred to as 'diffeomorphisms' (or 'general coordinate transformations'). The idea is that such deformations do not alter the physical situation at all. The principle of 'general covariance', which is a cornerstone of Einstein's general relativity, is that we formulate physical laws in such a way that such 'rubber-sheet deformations' (diffeomorphisms) do not alter the physically meaningful properties of the space and its contents.

This is not to say that all geometrical structure is lost, where the only kind of geometry that remains for our space might be something merely of the nature of its *topology* (indeed sometimes referred to as 'rubber-sheet geometry', in which the surface of a teacup would be identical to that of a ring, etc.). But we must be careful to specify what structure is needed. The term *manifold* is frequently used for such a space, of some definite finite number of dimensions (where we may refer to a manifold of n dimensions as an n-manifold), a manifold being *smooth* but not necessarily assigned any further structure beyond its smoothness and topology. In the case of hyperbolic geometry, there is actually a notion of *metric* assigned to the manifold—a mathematical 'tensor' quantity (see also §2.6), usually denoted by the letter **g**—which may be thought of as providing an assignment of a *length*[2.30] to any finite smooth curve in the space. Any deformation of the 'rubber sheet' constituting this manifold would carry with it any curve $\mathcal{C}$ connecting a pair of points p, q (where p and q are also carried by the deformation) and the length of the segment of $\mathcal{C}$ joining p to q assigned by **g** is deemed to be unaffected by this deformation (and, in this sense, **g** is also 'carried around' by the deformation).

This length notion also implies a notion of *straight line*, referred to as a *geodesic*, such a line l being characterized by the fact that for any two points p and q on l, not too far apart, the *shortest curve* (in the sense of length provided by **g**) from p to q is in fact the portion pq of l. See Fig. 2.15. (In this sense, a geodesic provides the 'shortest route between two points'.) We can also define *angles* between two smooth curves (this also being determined once **g** is given), so that the ordinary notions of geometry are available to us once **g** has been assigned. Nevertheless, this geometry would usually differ from the familiar Euclidean geometry.

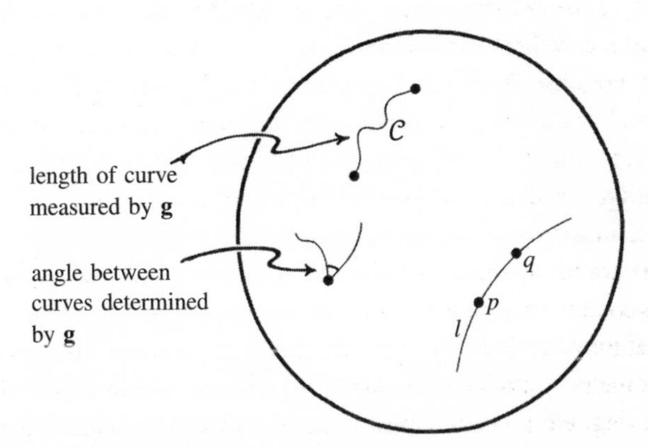

Fig. 2.15 The metric **g** assigns lengths to curves and angles between them. The geodesic l provides the 'shortest route between p and q' in the metric **g**.

The hyperbolic geometry of Escher's picture (Fig. 2.3(c), Beltrami–Poincaré conformal representation) thus also has its straight lines (geodesics). These can be understood in terms of the background

Euclidean geometry in which this figure is represented, as circular arcs meeting the boundary circle at right angles (see Fig. 2.16). Taking a and b to be the endpoints of the arc through two given points p and q, the hyperbolic **g**-*distance* between p and q turns out to be

$$C \log \frac{|qa||pb|}{|qb||pa|},$$

where the 'log' used here is a *natural* logarithm (2.302585. . . times the '$\log_{10}$' of §1.2), '$|qa|$' etc. being the ordinary Euclidean distances in the background space, and C is a positive constant called the *pseudo-radius* of the hyperbolic space.

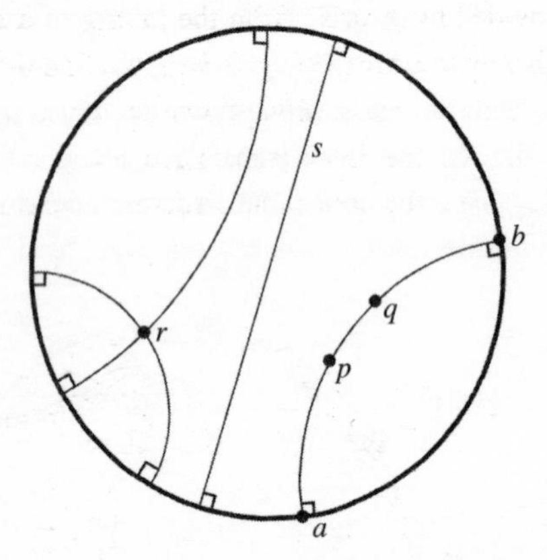

Fig. 2.16 'Straight lines' (geodesics) in conformal representation of hyperbolic geometry are circular arcs meeting the boundary circle at right angles.

But rather than specifying the structure provided by such a **g**, one may assign some other type of geometry instead. The kind that will be of most concern for us here is the geometry known as *conformal* geometry. This is the structure that provides a measure to the *angle* between two smooth curves, at any point where they meet, but a notion of 'distance' or 'length' is *not* specified. As mentioned above, the concept of angle is actually determined by **g**, but **g** itself is *not* fixed by the angle notion. While the conformal structure does not fix the length measure, it does fix the *ratios* of the length measures in different directions at any point—

so it determines infinitesimal *shapes*. We can rescale this length measure up or down at different points without affecting the conformal structure (see Fig. 2.17). We express this rescaling as

$$\mathbf{g} \mapsto \Omega^2\, \mathbf{g}$$

where Ω is a positive real number defined at each point, which varies smoothly over the space. Thus $\mathbf{g}$ and $\Omega^2\mathbf{g}$ give us the same conformal structure whatever positive Ω we choose, but $\mathbf{g}$ and $\Omega^2\mathbf{g}$ give us different metric structures (if $\Omega \neq 1$), where Ω is the factor of the scale change. (The reason for Ω appearing in 'squared' form in the expression '$\Omega^2\mathbf{g}$' is that the expressions for the direct measures of spatial—or temporal—separation, as provided by $\mathbf{g}$, arise from the taking of a *square root* (see Note 2.30).) Returning to Escher's Fig. 2.3(c), we find that the conformal structure of the hyperbolic plane (though *not* its metric structure) is actually identical to that of the Euclidean space interior to the bounding circle (yet differing from the conformal structure of the *entire* Euclidean plane).

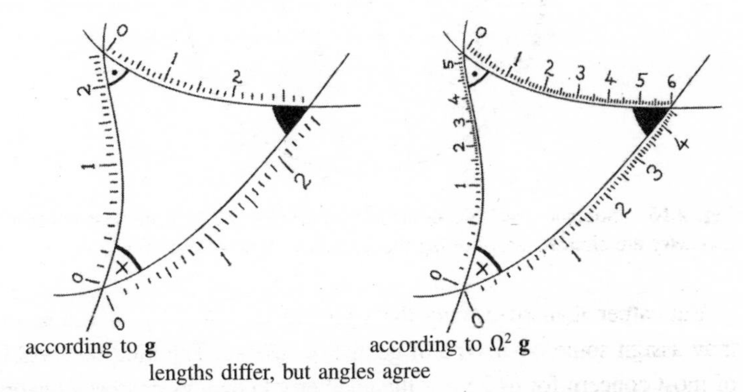

Fig. 2.17 Conformal structure does not fix length measure, but it does fix angles via the ratio of length measures in different directions at any point. Length measure can be rescaled up or down at different points without affecting the conformal structure.

When we come to *space-time* geometry, these ideas still apply, but there are some significant differences, owing to the 'twist' that Minkowski introduced into the ideas of Euclidean geometry. This twist is what mathematicians refer to as a change of the *signature* of the metric. In algebraic terms, this simply refers to a few + signs being changed to − signs, and it basically tells us how many of a set of n mutually orthogonal directions, for an n-dimensional space, are to be considered as 'timelike' (within the null cone) and how many 'spacelike' (outside the null cone). In Euclidean geometry, and its curved version known as *Riemannian* geometry, we think of *all* directions as being spacelike. The usual idea of 'space-time' involves only 1 of the directions being timelike, in such an orthogonal set, the rest being spacelike. We call it *Minkowskian* if it is flat and *Lorentzian* if it is curved. In the ordinary type of (Lorentzian) space-time that we are considering here, $n=4$, and the signature is '1+3' separating our 4 mutually orthogonal directions into 1 timelike direction and 3 spacelike ones. 'Orthogonality' between spacelike directions (and between timelike ones, had we had more than 1 of them) means simply 'at right angles', whereas between a spacelike and a timelike direction it looks geometrically more like the situation depicted in Fig. 2.18, the orthogonal directions being symmetrically related to the null direction between them. *Physically*, an observer whose world-line is in the timelike direction regards events in an orthogonal spacelike direction to be *simultaneous*.

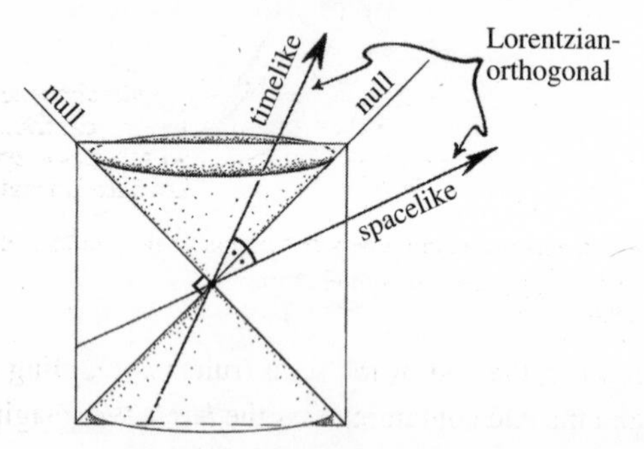

Fig. 2.18 'Orthogonality' of spacelike and timelike directions in Lorentzian space-time, represented in a Euclidean picture for which the null cone is right-angled.

In ordinary (Euclidean or Riemannian) geometry, we tend to think of lengths in terms of spatial separation, which is something that we might perhaps use a *ruler* to measure. But what is a ruler, in (Minkowskian or Lorentzian) space-time terms? It is a *strip*, which is not immediately the most obvious gadget for measuring the spatial separation between two events p and q. See Fig. 2.19. We can put p on one edge of the strip and q on the other. We can also assume that the ruler is narrow and unaccelerated, so that the space-time curvature effects of Einstein's (Lorentzian) general relativity are not of relevance, and a treatment according to special relativity should be adequate. But according to special relativity theory, in order for the distance measure provided by the ruler to give the correct space-time separation between p and q, we require that these events be *simultaneous* in the rest-frame of the ruler. How can we ensure that these events are actually simultaneous in the ruler's rest-frame? Well, we can use Einstein's original type of argument for this, although he was thinking more in terms of a *train* in uniform motion, than a ruler—so let us now phrase things that way too.

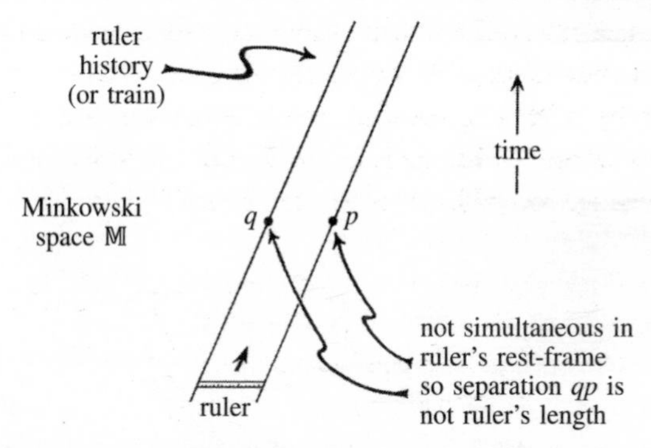

Fig. 2.19 A spacelike separation between points p and q in $\mathbb{M}$ is not directly measured by a ruler that is a 2-dimensional strip.

Let us refer to the end of the train (ruler) containing the event p as the *front*, and the end containing q as the *back*. We imagine an observer situated at the front, sending a light signal from an event r to the back of the train, timed so as to arrive there precisely at the event q, whereupon the signal is immediately reflected back to the front, to be received by the observer at the event s. See Fig. 2.20. The observer then judges q to be simultaneous with p, in the train's rest-frame, if

p occurs half-way between emission and final reception of the signal, i.e. if the time interval from r to p is precisely the same as that from p to s. The length of the train (i.e. of the ruler) then (and only then) would agree with the spatial interval between p and q.

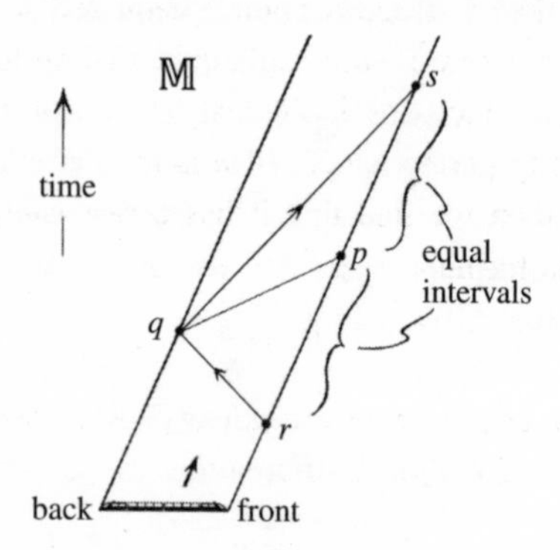

Fig. 2.20 The ruler (or train) measures the separation pq only when they are simultaneous, so light signals and clocks are needed instead.

We notice that not only is this a little more complicated than simply 'laying down a ruler' to measure the spatial separation between events, but what is actually measured by the observer would be the *time* intervals rp and ps. These (equal) time intervals *directly* provide the measure of the spatial interval pq that is being ascertained (in units where the speed of light c is taken to be unity). This illustrates the key fact about the metric of space-time, namely that it is really something that has much more directly to do with the measurement of *time* rather than distance. Instead of providing a *length* measurement for curves, it directly provides us with a *time* measurement. Moreover, it is not *all* curves that are assigned a time measure: it is for the curves referred to as *causal*, that could be the world-lines of particles, these curves being everywhere either *timelike* (with tangent vectors within the null cones, achieved by massive particles) or *null* (with tangent vectors along the null cones, achieved by massless particles). What the space-time metric **g** does is to assign a time measure to any finite segment of a causal curve (the contribution to the time measure being zero for any portion of the curve which is null). In this sense, the 'geometry' that the metric of space-time possesses should really be called 'chronometry', as the distinguished Irish relativity theorist John L. Synge has suggested.[2.31]

It is important for the physical basis of general relativity theory that extremely precise clocks actually exist in Nature, at a fundamental level, since the whole theory depends upon a naturally defined metric **g**.[2.32] In fact, this time measure is something quite central to physics, for there is a clear sense in which any individual (stable) massive particle plays a role as a virtually perfect clock. If m is the particle's mass (assumed to be constant), then we find that it has a *rest energy*[2.33] E given by Einstein's famous formula

$$E = mc^2,$$

which is fundamental to relativity theory. The other, almost equally famous formula—fundamental to *quantum* theory—is Max Planck's

$$E = h\nu$$

(h being Planck's constant), telling us that this particle's rest energy defines for it a particular frequency ν of quantum oscillation (see Fig. 2.21). In other words, any stable massive particle behaves as a very precise quantum *clock*, which 'ticks away' with the specific frequency

$$\nu = m\left(\frac{c^2}{h}\right),$$

in exact proportion to its mass, via the constant (fundamental) quantity c^2/h.

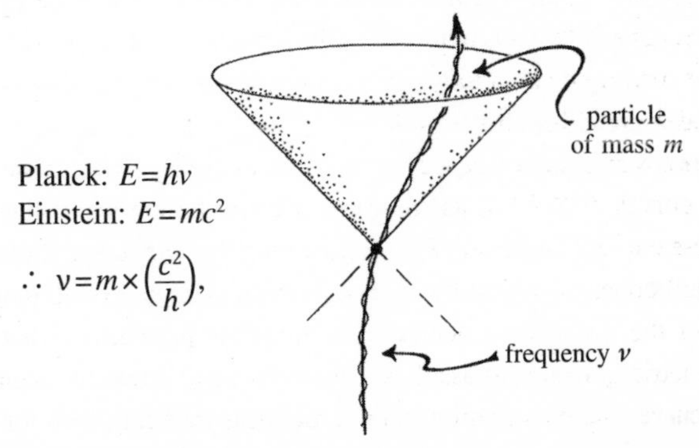

Fig. 2.21 Any stable massive particle behaves as a very precise quantum clock.

In fact the quantum frequency of a single particle is extremely high, and it cannot be directly harnessed to make a usable clock. For a clock that can be used in practice, we need a system containing many particles, combined together and acting appropriately in concert. But the key point is still that to build a clock we do need *mass*. Massless particles (e.g. photons) alone cannot be used to make a clock, because their frequencies would have to be *zero*; a photon would take until *eternity* before its internal 'clock' gets even to its first 'tick'! This fact will be of great significance for us later.

All this is in accordance with Fig. 2.22, where we see different identical clocks, all originating at the same event p, but moving with different velocities which are allowed to be comparable with (but less than) the speed of light. The bowl-shaped 3-surfaces (*hyperboloids*, in ordinary geometry) mark off the successive 'ticks' of the identical clocks. (These 3-surfaces are analogues of *spheres* for Minkowski's geometry, being the surfaces of constant 'distance' from a fixed point.) We note that a massless particle, since its world-line runs *along* the light cone, never reaches even the first of the bowl-shaped surfaces, in agreement with what has been said above.

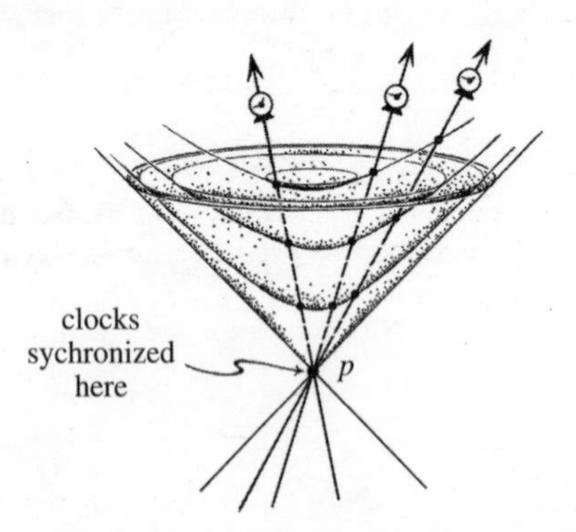

Fig. 2.22 Bowl-shaped 3-surfaces mark off the successive 'ticks' of identical clocks.

Finally, the notion of a *geodesic*, for a timelike curve, has the physical interpretation as the world-line of a massive particle in *free motion under gravity*. Mathematically, a timelike geodesic line l is characterized by the fact that for any two points p and q on l, not too far apart, the *longest curve* (in the sense of length of *time* provided by $\mathbf{g}$) from p to q is in fact a portion of l. See Fig. 2.23—a curious reversal of the length-*minimizing* property of geodesic Euclidean or Riemannian spaces. This notion of geodesic applies also to *null* geodesics, the 'length' being zero in this case, and the null-cone structure of the space-time alone is sufficient to determine them. This null-cone structure is actually equivalent to the space-time's *conformal* structure, a fact that will have importance for us later.

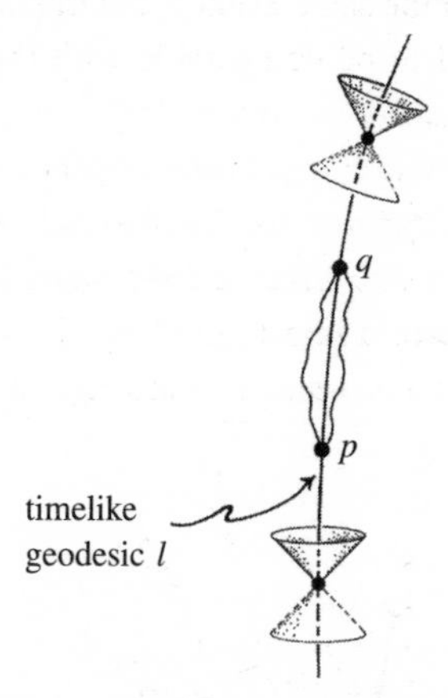

Fig. 2.23 A timelike geodesic line l is characterized by the fact that for any two points p and q on l, not too far apart, the *longest* local curve from p to q is in fact a portion of l.

2.4 Black holes and space-time singularities

In most physical situations, where the effects of gravity are comparatively small, the null cones deviate only slightly from their locations in Minkowski space $\mathbb{M}$. However, for a *black hole*, we find a very different situation, as I have tried to indicate in Fig. 2.24. This space-time picture represents the collapse of an over-massive star (perhaps ten, or more, times the mass of our Sun) which, having exhausted its resources of internal (nuclear) energy, collapses unstoppably inwards. At a certain stage—which may be identified as when the escape velolcity[2.34] from the star's surface reaches the speed of light—the inward tilt of the null cones becomes so extreme that the outermost part of the future cone becomes vertical in the diagram. Tracing out the envelope of these particular cones, we locate the 3-surface known as the *event horizon*, into which the body of the star finds itself to be falling. (Of course, I have had to suppress one of the spatial dimensions, in drawing this picture, so the horizon appears as an ordinary 2-surface, but this should not confuse the reader.)

Because of this tilt in the null cones, we find that any particle's world-line or light signal originating inside the event horizon will not be able to escape to the outside, as it would have to violate the requirements of §2.3 in order to cross the horizon. Also, if we trace back (in time) a light ray that enters the eye of an external observer, situated at a safe distance from the hole, looking towards it, we find that this ray cannot pass backwards across the event horizon into its interior, but hovers just above the surface, to meet the body of the star just a moment before it plunged beneath the horizon. This would theoretically be the case no matter how long the external observer waits (i.e. no matter how far up the picture we place the observer's eye), but in practice the image perceived by the observer would become highly red-shifted and very rapidly fade from view the later in time the observer is situated, so that in short order the image of the star would become blackness—in accordance with the terminology 'black hole'.

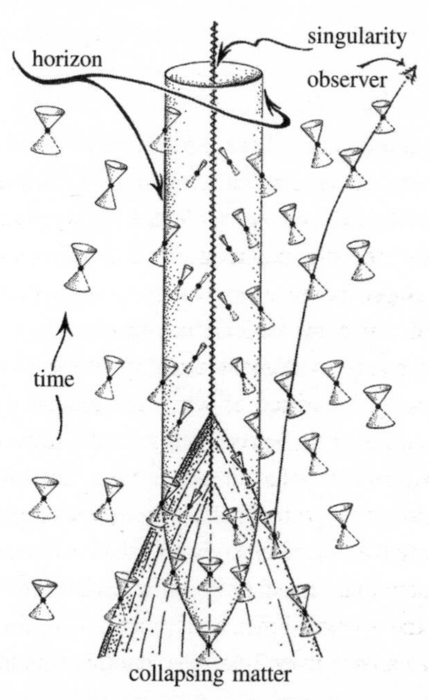

Fig. 2.24 Collapse of an over-massive star to a black hole. When the inward tilt of the future cone becomes vertical in the picture, light from the star can no longer escape its gravity. The envelope of these cones is the event horizon.

A natural question to ask is: what is to be the fate of this inward falling mass of material in the star after it crosses the horizon? Might it possibly indulge in some subsequent complicated activity, perhaps with the material swirling around, when it reaches the vicinity of the centre, effectively leading to an outward bounce? The original model of such a collapse, like that of Fig. 2.24, was put forward by J. Robert Oppenheimer and his student Hartland Snyder in 1939, and it was presented as an exact solution of the Einstein equations. However, various simplifying assumptions had to be made in order that they could represent their solution in an explicit way. The most important (and restrictive) of these was that exact *spherical symmetry* had to be assumed, so that such an asymmetrical 'swirling' could not be represented. They also assumed that the nature of the material of the star could be reasonably well approximated as a *pressureless fluid*—which is referred to by relativity theorists as 'dust' (see also §2.1). What Oppenheimer and Snyder found, under these assumptions, is that the inward collapse simply continues until the density of the material becomes *infinite* at a point at the centre, and the accompanying space-time curvature accordingly also becomes infinite. This central point, in their solution—represented by the vertical wiggly line in the middle of Fig. 2.24—is therefore referred to as a *space-time singularity*, where Einstein's theory 'gives up', and standard physics presents us with no way of evolving the solution further.

The presence of such space-time singularities has presented physicists with a fundamental conundrum, often viewed as the converse problem to that of the Big-Bang origin to the universe. Whereas the Big Bang is seen as the beginning of time, the singularities in black holes present themselves as representing the *end* of time—at least as far as the fate of that material that has, at some stage, fallen into the hole is concerned. In this sense, we may regard the problem presented by black-hole singularities to be the *time-reverse* of that presented by the Big Bang.

It is indeed true that every causal curve that originates within the horizon, in the black-hole collapse picture of Fig. 2.24, when extended into the future as far as it will go, must terminate at the central singularity. Likewise, in any of the Friedmann models referred to in §2.1, every causal curve (in the entire model), if extended as far back into the past as it will go, must terminate (actually originate) at the Big-Bang singularity. It would therefore appear that—apart from the black-hole case being more local—the two situations are, in effect, time-reverses of one another. Yet, our considerations of the Second Law might well suggest to us that this cannot altogether be the case. The Big Bang must be something of extraordinarily low entropy, in comparison with the situation to be encountered in a black hole. And the difference between one and the time-reverse of the other must be a key issue for our considerations here.

Before we come (in §2.6) to the nature of this difference, an important preliminary issue must be faced. We must address the question of whether, or to what extent, we have reason to *trust* the models—that of Oppenheimer and Snyder, on the one hand, and the highly symmetrical cosmological models such as that of Friedmann, on the other. We must note two of the significant assumptions underlying the Oppenheimer–Snyder picture of gravitational collapse. These are the *spherical symmetry* and the particular idealization of the material constituting the collapsing body that is taken to be completely *pressure free*. These two assumptions apply also to the Friedmann cosmological models (the spherical symmetry applying to *all* FLRW models), so we may well have cause to doubt that these idealized models need necessarily represent the inevitable behaviour of collapsing (or exploding) matter, in such extreme situations, according to Einstein's general relativity.

In fact, both these issues were matters that concerned me when I started thinking seriously about gravitational collapse in the autumn of 1964. This had been stimulated by concerns expressed to me by the deeply insightful American physicist John A. Wheeler, following the recent discovery, by Maarten Schmidt, of a remarkable object[2.35] whose extraordinary brightness and variability indicated that something approaching the nature of what we now call a 'black hole' might have to be involved. At that time, there had been a common belief, based on some detailed theoretical work that had been carried out by two Russian physicists, Evgeny Mikhailovich Lifshitz and Isaak Markovich Khalatnikov, that in the *general* situation, where no conditions of symmetry would apply, space-time singularities would *not* arise in a general gravitational collapse. Being only vaguely aware of the Russian work, but having my doubts that the kind of mathematical analysis that they had been employing would be likely to lead to any definitive conclusion on this matter, I started thinking about the problem in my own rather more geometrical way. This involved my trying to understand various global aspects of how light rays propagate, how they may be focused by space-time curvature, and what kind of singular surfaces might arise when they start to crinkle and cross over one another.

I had earlier been thinking in these terms in relation to the steady-state model of the universe, referred to at the beginning of §2.2. Having been quite fond of that model, but not so fond of it as I had been of Einstein's general relativity—with its magnificent unification of basic space-time geometrical notions with fundamental physical principles—I had wondered whether there might be any possibility that the two could be made to be consistent with one another. If one sticks to the pure smoothed-out steady-state model, one is rapidly forced to conclude that this consistency cannot be achieved without the introduction of *negative energy densities*, these having the effect, in Einstein's theory, of being able to spew light rays apart, in order to counter the relentless inward curving effect of the positive energy density of normal matter (see §2.6). In a general way, the presence of negative energy in physical systems is 'bad news', as it is likely to lead to uncontrollable instabilities. So I wondered whether deviations from symmetry might allow one to avoid such unpleasant conclusions. However, the global arguments that can be used to address the topological behaviour of such light-ray surfaces turn out to be so powerful, if due care is exercised, that they can often be applied in quite general situations to derive the same sort of conclusion as applies when this high symmetry *is* assumed. The upshot was (though I never published these conclusions) that reasonable departures from symmetry do not really help, and so the steady-state model, even when considerable deviations from the symmetrical smoothed-out model are allowed, cannot escape being inconsistent with general relativity unless negative energies are present.

I had also used some similar types of argument to investigate the different possibilities that may arise when one considers the remote future of gravitating systems. The techniques that I was led to, involving the ideas of conformal space-time geometry (referred to in §2.3 above and which will have important roles to play in Part 3), also led me to consider the focusing properties of light-ray systems[2.36] in general situations, so

I began to believe that I was fairly much at home with these things, and I turned my attention to the question of gravitational collapse. The main additional difficulty here was that one needed some kind of criterion to characterize situations in which the collapse had passed a 'point of no return', for there are many situations in which the collapse of a body can be overturned because pressure forces become large enough to reverse a collapse, so that the material 'bounces' out again. Such a point of no return seems to arise when the horizon forms, since gravitation has then become so strong that it has overcome everything else. However, the presence and location of a horizon turns out to be an awkward thing to specify mathematically, its precise definition actually requiring behaviour to be examined all the way out to infinity. Accordingly, it was fortunate for me that an idea occurred to me[2.37]—that of a 'trapped surface'—which was of a rather more local character,[2.38] whose presence in a space-time may be taken as a condition that an unstoppable gravitational collapse has indeed taken place.

By use of the type of 'light-ray/topology' argument that I had been developing I was then able to establish a theorem[2.39] to the effect that whenever such a gravitational collapse has taken place, singularities cannot be avoided, provided a couple of 'reasonable' conditions are satisfied by the space-time. One of these is that the light-ray focusing cannot ever be negative; in more physical terms this means that if Einstein's equations are assumed (with or without the presence of a cosmological constant Λ), the *energy flux* across a light ray is never negative. A second condition is that the whole system must be able to be evolved from an open (i.e. what is called 'non-compact') spacelike 3-surface Σ. This is a very standard situation for considering a reasonably localized (i.e. non-cosmological) physically evolving situation. Geometrically, all we require is that any causal curve in the space-time under consideration, to the future of Σ, when extended backwards (in time) as far as it will go, must intersect Σ (see Fig. 2.25). The only other requirement (apart from the assumed existence of a trapped surface) concerns what is actually to be meant by a 'singularity' in this context. Basically, a singularity simply represents an obstruction to continuing the space-time smoothly, indefinitely into the future,[2.40] consistently with the assumptions just made.

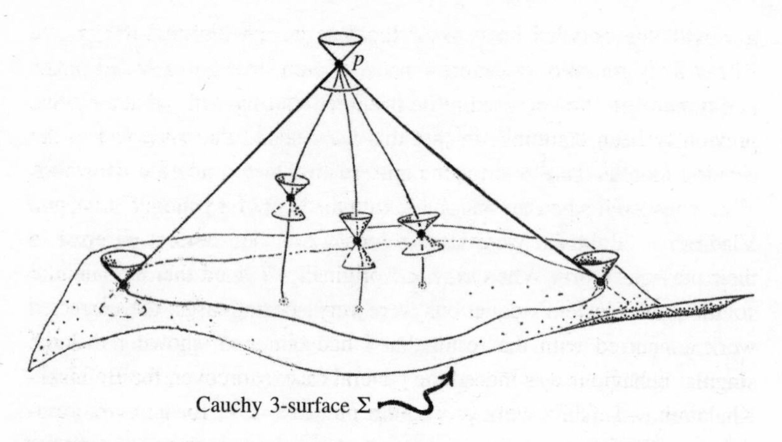

Fig. 2.25 An initial 'Cauchy surface' Σ; any point p to its future has the property that every causal curve terminating at p, when extended back far enough to the past, must meet the surface.

The power of this result lies in its generality. Not only is there no assumption of symmetry required, nor of any other simplifying condition that might make the equations easier to solve, but the nature of the material source of the gravitational field is constrained only to be 'physically reasonable' according to the physical requirement that the energy flux of this material across any light-ray must never be negative—a condition known as the 'weak energy condition'. This condition is certainly satisfied by the pressure-free dust assumed by Oppenheimer and Snyder, and also by Friedmann. But it is far more general than this, and includes every type of physically realistic classical material that is considered by relativity theorists.

Complementary to this strength, however, is the weakness of this result that it tells us almost nothing whatever about the detailed nature of the problem confronting our collapsing star. It gives no clue as to the geometrical form of the singularity. It does not even tell us that the matter will reach infinite density or that the space-time curvature will become infinite in any other way. Moreover, it tells us nothing even about *where* the singular behaviour will begin to show itself.

To address such matters, something is needed that is much more in line with the detailed analysis of the Russian physicists Lifshitz and Khalatnikov referred to above. Yet the theorem that I had found in the late months of 1964 seemed to be in direct conflict with what they had previously been claiming! In fact this was indeed the case, and in the ensuing months there was much consternation and confusion. However, all was resolved when the Russians, with the help of a younger colleague Vladimir A. Belinski, were able to locate and then correct an error in their previous work. Whereas it had originally seemed that the singular solutions of Einstein's equations were very special cases, the corrected work concurred with the result that I had obtained, showing that the singular behaviour was indeed the general case. Moreover, the Belinski–Khalatnikov–Lifshitz work provided a plausible case for an extraordinarily complicated chaotic type of activity for the approach to a singularity now referred to as the *BKL-conjecture*. Such behaviour had already been anticipated from considerations by the American relativity theorist Charles W. Misner—referred to as the *mixmaster* universe—and it seems to me to be quite possible that at least in a broad class of possible situations, such wild and chaotic 'mixmaster' activity is likely in the general case.

I shall have more to say about this matter later (in §2.6) but, for now, we must address another issue, namely whether something like the occurrence of a trapped surface is actually likely to arise in any plausible situation. The original reason for anticipating that over-massive stars might actually collapse catastrophically at a late stage of their evolution arose from the work of Subrahmanyan Chandrasekhar, in 1930, when he studied the structure of the miniature, hugely dense stars known as *white dwarfs* (the first known example being the mysterious dark companion of the bright star Sirius), of mass comparable with that of the Sun but with a radius roughly that of the Earth. White dwarfs are held apart by *electron degeneracy pressure*—a quantum-mechanical principle which prevents electrons from getting crowded on top of one another. Chandrasekhar showed that when the effects of (special) relativity are brought in, there is a limit to the mass that can sustain itself against gravity in this way, and he drew attention to the profound conundrum that this raises for cold masses *larger* than this 'Chandrasekhar limit'. This limit is about $1.4 M_\odot$ (where $M_\odot$ denotes the mass of the Sun).

The evolution of an ordinary ('main sequence') star like our Sun involves a late stage where its outer layers swell, so that it becomes a huge *red giant*, accompanied by an electron-degenerate core. This core gradually accumulates more and more of the star's material, and if this does not result in Chandrasekhar's limit being exceeded, the entire star can end up as a white dwarf, eventually cooling down to end its existence as a black dwarf. This, indeed, is the expected fate of our own Sun. But for much larger stars, the white-dwarf core could *collapse* at some stage, owing to Chandrasekhar's limit being exceeded, the infalling material in the star leading to an extremely violent *supernova* explosion (probably outshining, for a few days, the entire galaxy in which it resides). Sufficient material might be shed, during this process, so that the resulting core is able to be sustained at an even far greater density (with, say, $1.5\,M_\odot$ compressed into a region of around 10 km in diameter), forming a *neutron star*, which is sustained by *neutron degeneracy pressure*.

Neutron stars sometimes reveal themselves as *pulsars* (see §2.1 and note 2.6) and many have now been observed in our galaxy. But again there is a limit on the possible mass of such a star, this being around $1.5\,M_\odot$ (sometimes known as the *Landau limit*). If the original star had been sufficiently massive (say more than $10\,M_\odot$), then it is very likely that insufficient material would be blown off in the explosion, and the core would be unable to sustain itself as a neutron star. Then there is nothing left to stop its collapse, and in all probability a stage would be reached in which a trapped surface arises.

Of course, this is not a definitive conclusion, and one might well argue that not enough is known about the physics of such extraordinarily condensed states that the material would reach before the trapped-surface regime is reached (though only about a factor of 3 in the radius, down from that of a neutron star). However, the case for black holes arising is considerably stronger if we consider mass concentrations on the far larger scale of collections of many stars near galactic centres. This is simply a matter of how things scale. For larger and larger systems, trapped surfaces would arise at smaller and smaller densities. There is enough room, for example, for about a million white dwarf stars, none of which need be actually in contact, to occupy a region of 10^6 km in diameter, and this would be small enough for a trapped surface to arise surrounding them. The issue of 'unknown physics' at extremely high densities is not really the relevant one when it comes to the formation of black holes.

There is one further theoretical issue which I have glossed over so far. I have been tacitly assuming that the existence of a trapped surface implies that a black hole will form. This deduction, however, depends upon what is referred to as 'cosmic censorship' which, though widely believed to be true, remains an unproved conjecture,[2.41] as of now. Along with the BKL conjecture, it is probably the major unsolved issue of classical general relativity. What cosmic censorship asserts is that *naked* space-time singularities do not occur in generic gravitational collapses, where 'naked' means that causal curves originating at the singularity can escape to reach a distant external observer (so that the singularity is not shielded from external observation by an event horizon). I shall return to the issue of cosmic censorship in §2.6.

In any case, the *observational* situation at the present time very strongly favours the presence of black holes. The evidence that certain binary star systems contain black holes of a few solar masses is rather impressive, although it is of the somewhat 'negative' character that an invisible component to the system makes its presence evident from the dynamical motions, the mass of the invisible component being considerably larger than could be the case for any compact object, according to standard theory. The most impressive observations of this kind occur with the very rapid orbital motions of observed stars around an invisible but enormously massive compact entity at the centre of our Milky Way galaxy. The speed of these motions is such that this entity must have a mass of about 4 000 000 $M_\odot$! It is hard to imagine that this can be anything other than a black hole. In addition to evidence of this 'negative' kind, there are also entities of this nature that are observed to be dragging in surrounding material, where this material shows no evidence of heating up a 'surface' to the entity. The lack of a ponderable surface is clear direct evidence for a black hole.[2.42]

2.5 Conformal diagrams and conformal boundaries

There is a convenient way of representing space-time models in their entirety, especially in the case of models possessing spherical symmetry, as in the case of the Oppenheimer–Snyder and Friedmann space-times. This is by the use of *conformal diagrams*. I shall distinguish two types of conformal diagram here, the *strict* and the *schematic* conformal diagrams.[2.43] We shall be seeing something of the utility of each.

Let us start with the strict conformal diagrams, which can be used to represent space-times (here denoted by $\mathcal{M}$) with exact spherical symmetry. The diagram would be a region $\mathcal{D}$ of the plane, and each point in the interior of $\mathcal{D}$ would represent a whole sphere's worth (i.e. an S^2's worth) of points of $\mathcal{M}$. To get something of a picture of what is going on, we may lose one spatial dimension, and imagine *rotating* the region $\mathcal{D}$ about some vertical line off to the left (see Fig. 2.26)—this line being referred to as an *axis* of rotation. Then each point in $\mathcal{D}$ will trace out a *circle* (S^1). This is good enough for our visual imaginations. But for the full 4-dimensional picture of our space-time $\mathcal{M}$, we would need a *2-dimensional* rotation, so each interior point of $\mathcal{D}$ has to trace out a *sphere* (S^2) in $\mathcal{M}$.

Often, in our strict conformal diagrams, we find that we have an axis of rotation which is part of the *boundary* to the region $\mathcal{D}$. Then those boundary points on the axis—represented in the diagram as a *broken* line—would each represent a single *point* (rather than an S^2) in the 4-dimensional space-time, so that the entire broken line would

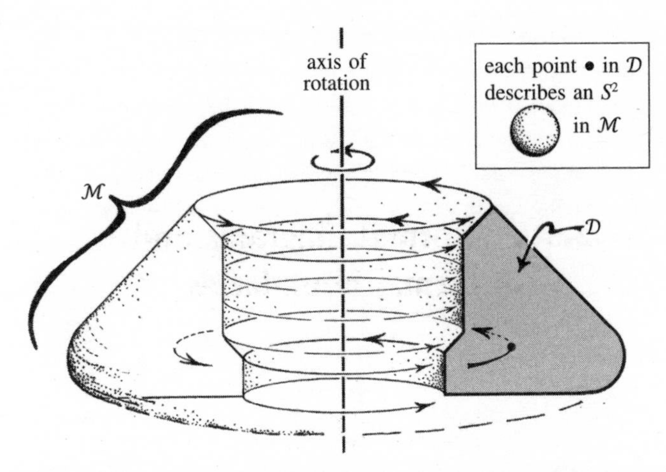

Fig. 2.26 Strict conformal diagram $\mathcal{D}$ used to represent space-times (here denoted by $\mathcal{M}$) with exact spherical symmetry. The 2-dimensional region $\mathcal{D}$ is rotated (through a 2-dimensional sphere S^2) to make the 4-space $\mathcal{M}$.

also represent a single line in $\mathcal{M}$. Fig. 2.27 gives us an impression of how the whole space-time $\mathcal{M}$ is constituted as a family of 2-dimensional spaces identical to $\mathcal{D}$ in rotation about the broken-line axis.

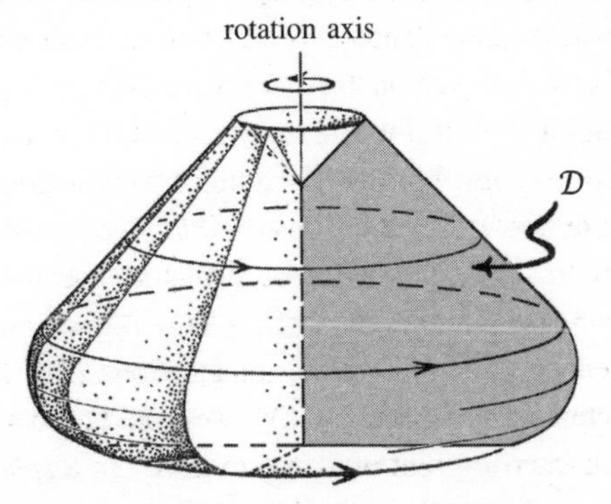

Fig. 2.27 The broken line on $\mathcal{D}$'s boundary is a symmetry axis, each point of which represents a single space-time point, rather than an S^2.

We are going to think of $\mathcal{M}$ as a *conformal* space-time, and not worry too much about the particular scaling that gives $\mathcal{M}$ its full metric **g**.

Thus, in accordance with the final sentence of §2.3, $\mathcal{M}$ is provided with a full family of (time-oriented) null cones. In accordance with this, $\mathcal{D}$ itself, being a 2-dimensional subspace of $\mathcal{M}$, inherits from it a 2-dimensional conformal space-time structure, and has its own 'time-oriented null cones'. These simply consist of a pair of distinct 'null' directions at each point of $\mathcal{D}$ that are deemed to be oriented towards the future. (They are just the intersections of the planes defining the copies of $\mathcal{D}$ with the future null cones of $\mathcal{M}$; see Fig. 2.28.)

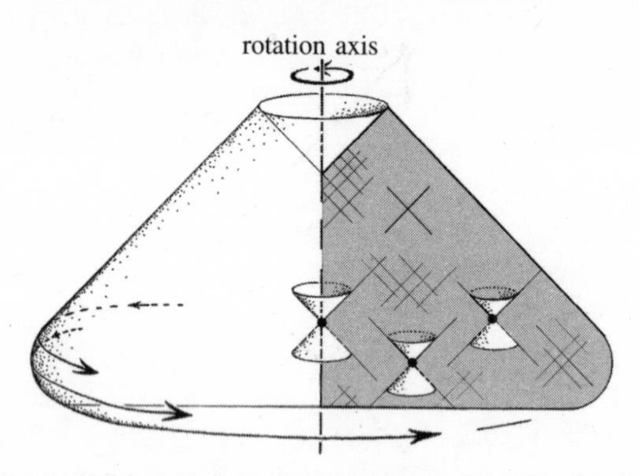

Fig. 2.28 The 'null cones' in $\mathcal{D}$, angled at 45° to the vertical, are the intersections of those in $\mathcal{M}$ with an embedded $\mathcal{D}$.

In a strict conformal diagram, we endeavour to arrange all these future null directions in $\mathcal{D}$ to be oriented at 45° to the upward vertical. To illuminate the situation, I have drawn, in Fig. 2.29, a conformal diagram for the entire Minkowski space-time $\mathbb{M}$, the radial null lines being drawn at 45° to the upward vertical. In Fig. 2.30, I have tried to indicate how this mapping is achieved. We see that Fig. 2.29 exhibits an important feature of conformal diagrams: the picture is of merely a *finite* (right-angled) triangle, despite the entire infinite space-time $\mathbb{M}$ being encompassed by the diagram. A characteristic feature of conformal diagrams is, indeed, that they enable the infinite regions of the space-time to be 'squashed down' so as to be encompassed by a finite picture. Infinity itself is also represented in the diagram. The two bold sloping boundary lines represent past null infinity $\mathscr{I}^-$ and future null infinity $\mathscr{I}^+$, where every null geodesic (null straight line) in $\mathbb{M}$ acquires a past end-point on $\mathscr{I}^-$ and a future end-point on $\mathscr{I}^+$. (It is usual to pronounce the letter $\mathscr{I}$ as 'scri'—meaning 'script I'.)[2.44] There are also three points, i^-, i^0, and i^+ on the boundary, respectively representing past timelike infinity, spacelike infinity, and future timelike infinity, where every timelike geodesic in $\mathbb{M}$ acquires the past end-point i^- and future end-point i^+, and every spacelike geodesic closes into a loop via the point i^0. (We shall be seeing, shortly, why i^0 must indeed be considered as being just a single point.)

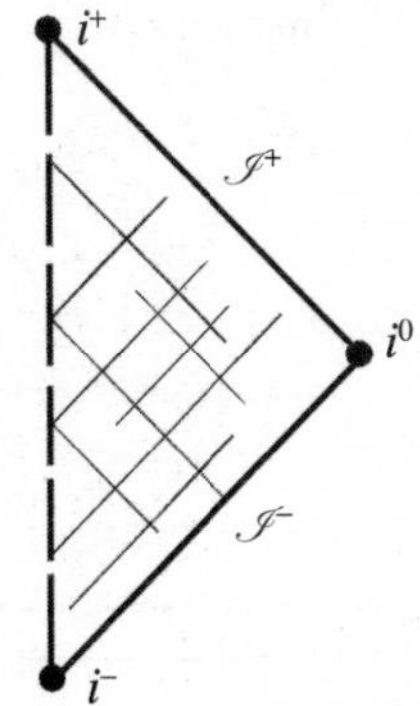

Fig. 2.29 Strict conformal diagram of Minkowski space $\mathbb{M}$.

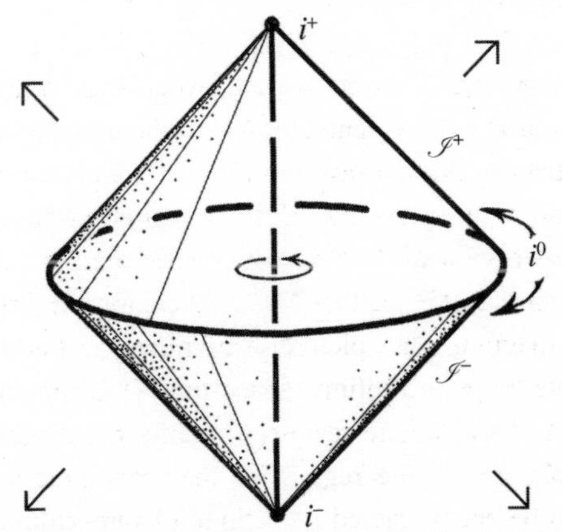

Fig. 2.30 To get to the normal picture of $\mathbb{M}$, imagine the sloping (conical) boundaries pushed outwards to infinity.

At this juncture, it may be helpful to recall the Escher print Fig. 2.3(c) providing a conformal picture of the entire hyperbolic plane. The bounding circle represents its *infinity*, in a conformally finite way, in an essentially similar manner to the way in which $\mathscr{I}^+$, $\mathscr{I}^-$, i^-, i^0, and i^+ together represent infinity for $\mathbb{M}$. In fact, just as we can *extend* the hyperbolic plane, as a smooth conformal manifold, beyond its conformal boundary to the Euclidean plane inside which it is represented (Fig. 2.31), we may also extend $\mathbb{M}$, smoothly, beyond its boundary to a larger conformal manifold. In fact, $\mathbb{M}$ is conformally identical to a portion of the space-time model known as the *Einstein universe* $\mathcal{E}$ (or the 'Einstein cylinder'). This is a cosmological model which is spatially a 3-sphere (S^3) and completely static. Figure 2.32(a) gives an intuitive picture of this model (the one Einstein originally introduced his cosmological constant Λ in order to achieve, in 1917; see §2.1) and Fig. 2.32(b) provides a strict conformal diagram representing it. Note that in this diagram there are two separate 'axes of rotation', represented by the two vertical broken lines. This is completely consistent; we just think of the radius of the S^2, which each point in the interior of the diagram represents, as shrinking down to zero as a broken line is approached. This also serves to explain the rather

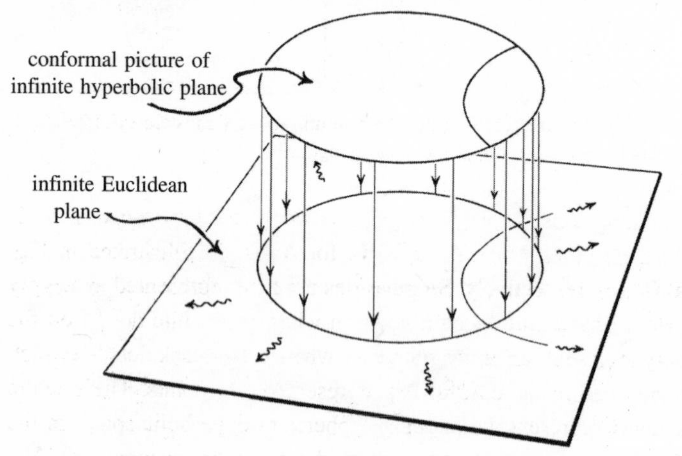

Fig. 2.31 Extending the hyperbolic plane, as a smooth conformal manifold, beyond its conformal boundary to the Euclidean plane inside which it is represented.

curious-seeming fact that spatial infinity for $\mathbb{M}$ is conformally just the *single point* i^0, for the radius of the S^2 that it would seem to have represented has shrunk down to zero. The spatial S^3 cross-sections of the space-time $\mathcal{E}$ arise from this procedure. Figure 2.33(a) shows how $\mathbb{M}$ arises as a conformal subregion of $\mathcal{E}$, and in fact how we can consider the entire manifold $\mathcal{E}$ as made up, conformally, of an infinite succession of spaces $\mathbb{M}$, where the $\mathscr{I}^+$ of each one is joined on to the $\mathscr{I}^-$, of the next, and Fig. 2.33(b) shows how this is done in terms of strict conformal diagrams. It will be worth bearing this picture in mind when we come to consider the proposed model of Part 3.

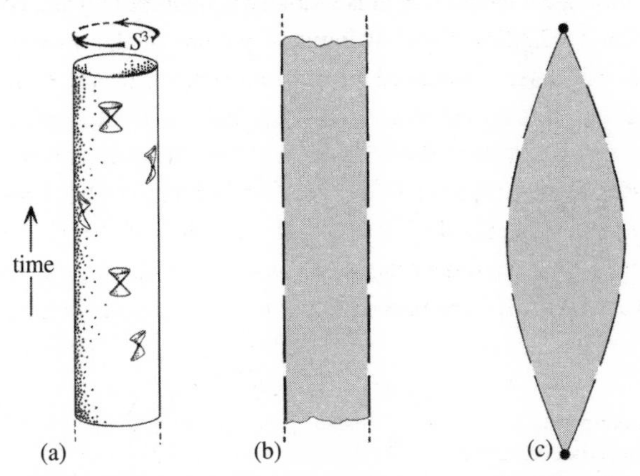

Fig. 2.32 (a) Intuitive picture of the Einstein universe $\mathcal{E}$ ('Einstein cylinder'); (b), (c) strict conformal diagrams of the same thing.

Let us now consider the Friedmann cosmologies introduced in §2.1. The different cases $K>0$, $K=0$, $K<0$, for $\Lambda=0$, are illustrated in Fig. 2.34(a),(b),(c), respectively. Singularities are here represented as wiggly lines. Here I have introduced a notation whereby a white dot '∘' on the boundary represents an entire sphere S^2, whereas the black dots '•' (which we already had in the case of $\mathbb{M}$) represent single points. These white dots actually represent the boundary spheres of hyperbolic space, in the conformal representation that Escher used in the 2-dimensional case. The corresponding cases for positive cosmological constant ($\Lambda>0$, where in

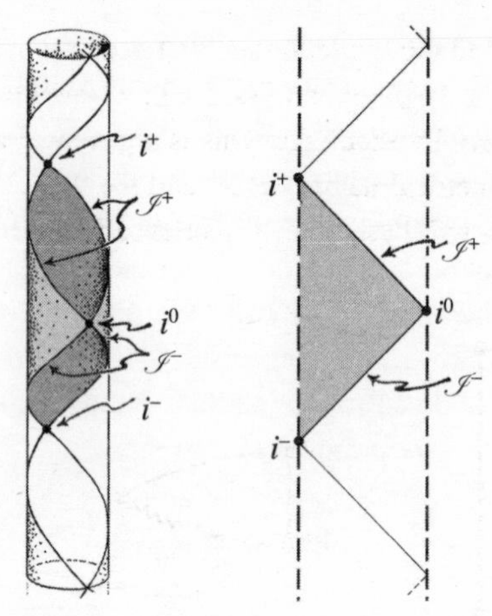

Fig. 2.33 To see why i^0 is a single point. (a) $\mathbb{M}$ arises as a conformal subregion of $\mathcal{E}$. The entire manifold $\mathcal{E}$ can be considered to be made up, conformally, of an infinite succession of spaces $\mathbb{M}$; (b) shows how this is done in terms of strict conformal diagrams.

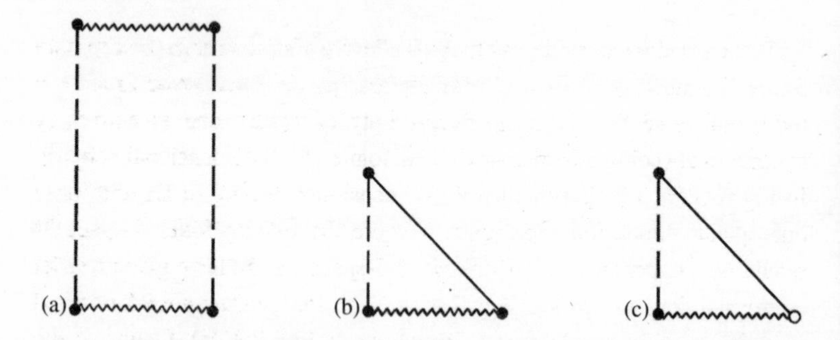

Fig. 2.34 Strict conformal diagrams for the three different cases, $K>0$, $K=0$, $K<0$ for $\Lambda=0$, of the Friedmann cosmologies.

the case $K>0$ we assume that the spatial curvature is not large enough to overcome Λ and produce an ultimate re-collapse) are illustrated in Fig. 2.35(a),(b),(c). An important feature of these diagrams may be pointed out here. The future infinity $\mathscr{I}^+$ of all these models is *spacelike*, as is indicated by the final bold boundary line being always more horizontal than

45°, in contrast with the future infinity that occurs when $\Lambda=0$ (in the cases illustrated in Fig. 2.34(b),(c) and Fig. 2.29), where the boundary is at 45°, so $\mathscr{I}^+$ is then a *null* hypersurface. This is a general feature of the relation between the geometrical nature of $\mathscr{I}^+$ and the value of the cosmological constant Λ, and it will have a key importance for us in Part 3.

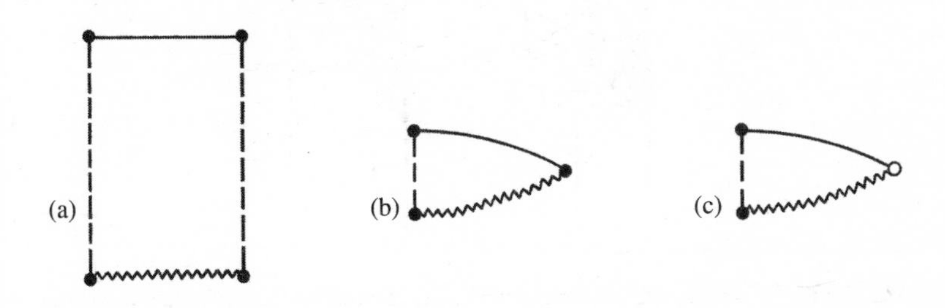

Fig. 2.35 Strict conformal diagrams for Friedmann models with $\Lambda>0$. (a) $K>0$; (b) $K=0$; (c) $K<0$.

These Friedmann models with $\Lambda>0$ all have a behaviour in their remote future (i.e. near $\mathscr{I}^+$) which closely approaches *de Sitter* space-time $\mathbb{D}$, a model universe that is completely empty of matter and is extremely symmetrical (being a Minkowskian analogue of a 4-dimensional sphere). In Fig. 2.36(a) I have sketched a 2-dimensional version of $\mathbb{D}$, with only one spatial dimension represented (where the full de Sitter 4-space $\mathbb{D}$ would be a hypersurface in Minkowski 5-space), and I have given a strict conformal diagram for it in Fig. 2.36(b). The *steady-state model*, referred to in §2.2, is just one half of $\mathbb{D}$, as shown in Fig. 2.36(c). Owing to the 'cut' through $\mathbb{D}$ that is required (jagged boundary), the steady-state model is actually what is called 'incomplete', in past directions. There are ordinary timelike geodesics—which could represent free motions of massive particles—whose time measure does not extend to earlier values than some finite value. This might well have been regarded as a worrying flaw in the model if it had applied to future directions, since it could apply to the future of some particle or space traveller,[2.45] but here we can simply say that such particle motions were never present.

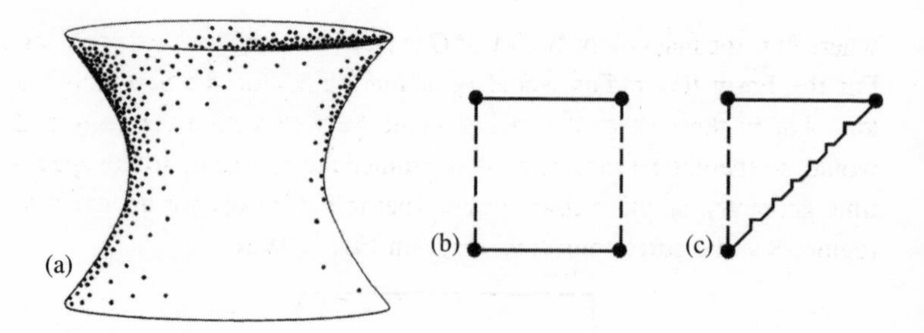

Fig. 2.36 De Sitter space-time: (a) represented (with 2 spatial dimensions suppressed) in Minkowski 3-space; (b) its strict conformal diagram; (c) cut in half, we get a strict conformal diagram for the steady-state model.

Whatever view one might take on the physics of the matter, I indicate this kind of incompleteness by a slightly jagged line in my strict conformal diagrams. The one remaining type of line that I am using in these diagrams is an internal dotted line, to denote a black hole's event horizon. I am using all these five kinds of line (broken for symmetry axis, bold for infinity, wiggly for a singularity, slightly jagged for incompleteness, and dotted for a black hole's horizon) and two kinds of spot (black representing a single point in the 4-space, white tracing out an S^2) consistently in my strict conformal diagrams, as given in the key in Fig. 2.37.

A strict conformal diagram for the Oppenheimer–Snyder collapse to a black hole is given in Fig. 2.38(a). This arises from 'gluing together' a portion of a collapsing Friedmann model and a portion of the Eddington–Finkelstein extension of the original Schwarzschild solution, as shown in the strict conformal diagrams Fig. 2.38(b),(c); see also Fig. 2.39. Schwarzschild found his solution of Einstein's equations in 1916, shortly after Einstein published the equations for his general theory of relativity. This solution describes the external gravitational field of a static spherically-symmetrical body (such as a star), and it can be extended inwards, as a static space-time, down to its *Schwarzschild radius*

$$\frac{2MG}{c^2}$$

where M is the mass of the body and G is Newton's gravitational constant. For the Earth this radius would be about 9 mm, for the Sun, about 3 km—but in these cases the radius would be well within the body and would be theoretical distances of no immediate relevance to the space-time geometry, as this Schwarzschild metric holds only for the external region. See the strict conformal diagram Fig. 2.39(a).

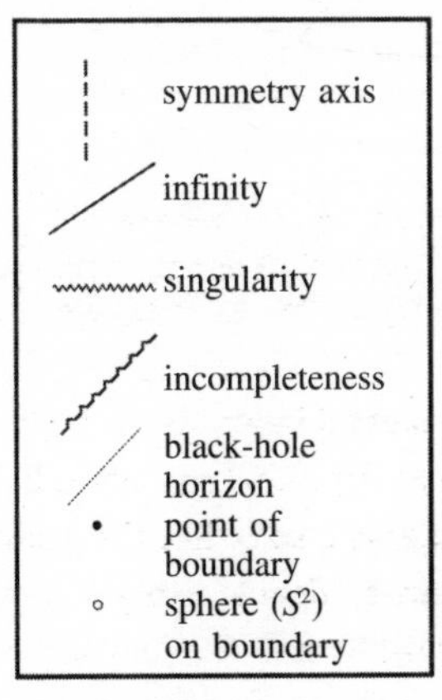

Fig. 2.37 Key for strict conformal diagrams.

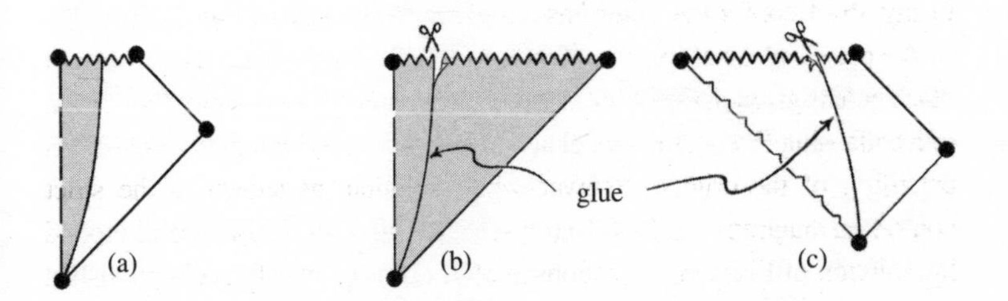

Fig. 2.38 The Oppenheimer–Snyder model of collapse to a black hole: (a) strict conformal diagram constructed from gluing together; (b) left part of time-reverse of Friedmann model (Fig. 2.34(b)) and (c) right part of Eddington–Finkelstein model (Fig. 2.39(b)). (In local models, such as these, Λ is ignored, so $\mathcal{J}$ is treated as null.)

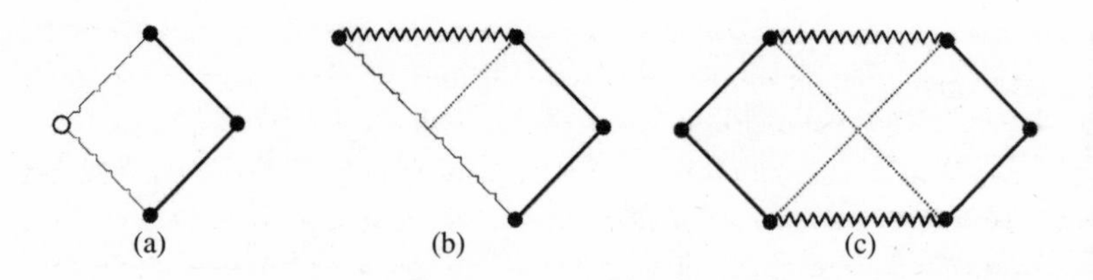

Fig. 2.39 Strict conformal diagrams of a spherically symmetrical ($\Lambda = 0$) vacuum: (a) original Schwarzschild solution, external to the Schwarzschild radius; (b) extension to Eddington–Finkelstein collapse metric; (c) full extension to Kruskal/Synge/Szekeres/Fronsdal form.

For a black hole, however, the Schwarzschild radius would be at the horizon. At this radius, the Schwarzschild form of the metric goes singular, and the Schwarzschild radius was originally thought of as an actual singularity in space-time. However it was found, initially by Georges Lemaître in 1927, that if we abandon the requirement that the space-time remain static, it is possible to extend it in a completely smooth way. A simpler description of this extension was found by Arthur Eddington in 1930 (although he omitted to point out what it had achieved in this respect), this description being rediscovered, and its implications clearly enunciated, by David Finkelstein in 1958; see the strict conformal diagram of this in Fig. 2.39(b). What is referred to as the 'maximal extension of the Schwarzschild solution' (often called the Kruskal–Szekeres extension, although an equivalent, though more complicated description was found much earlier by J.L. Synge[2.46]) is given in the strict conformal diagram of Fig. 2.39(c).

In §3.4, we shall come to another feature of black holes which, though an extremely tiny effect at the present time, will ultimately have a crucial significance for us. Whereas according to the classical physics of Einstein's general relativity a black hole ought to be completely black, an analysis carried out by Stephen Hawking in 1974 showed that when effects of quantum field theory in curved space-time backgrounds are brought into the picture, a black hole ought to have a very tiny temperature T, which is inversely proportional to the hole's mass. For a black hole of $10\,M_\odot$, for example, this temperature would be the extraordinarily tiny, around 6×10^{-9} K which may be compared with the record low temperature, as of 2006, produced in the laboratory of $\sim10^{-9}$ K achieved at MIT. This is about as warm as the black holes around today are likely to be. Larger black holes would be even colder, and the temperature of the $\sim 4\,000\,000 M_\odot$ black hole at the centre of our galaxy would be only about 1.5×10^{-14} K. Taking the ambient temperature of our universe, at the present time to be that of the CMB, we find that it has the immensely hotter value of ~ 2.7 K.

Yet, if we take the *very* very long view, and bear in mind that the exponential expansion of our universe will, if it continues indefinitely, lead to a vast cooling in the CMB, we would expect it to get down to the temperature of even the largest black holes that are likely ever to arise. After that, the black hole will start to radiate away its energy into the surrounding space, and in losing energy it must also lose mass (by Einstein's $E=mc^2$). As it loses mass, it will get hotter, and gradually, after an incredible length of time (perhaps up to around 10^{100}—i.e. a 'googol'—years, for the largest black holes around today) it shrinks away completely, finally disappearing with a 'pop'—this final explosion being hardly worthy of the name 'bang', as it would be likely to be only around the energy of an artillery shell which is something of an anticlimax after all that wait!

Of course, this is extrapolating our present physical knowledge and understanding to an enormous degree. But Hawking's analysis is well in accordance with accepted general principles, and these principles do not seem to allow us to escape this overall conclusion. Accordingly, I am accepting it as a plausible account of a black hole's eventual fate. Indeed, this expectation will form an important ingredient to the scheme that I shall be presenting in Part 3 of this book. In any case, it is of relevance to present a sketch of this process in Fig. 2.40, together with its strict conformal diagram, in Fig. 2.41.

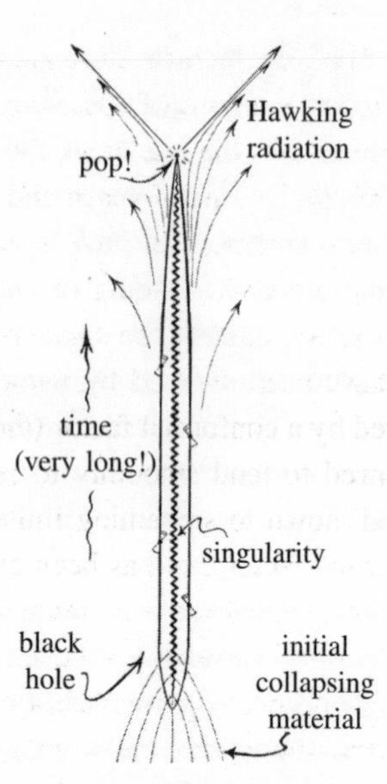

Fig. 2.40 Hawking-evaporating black hole.

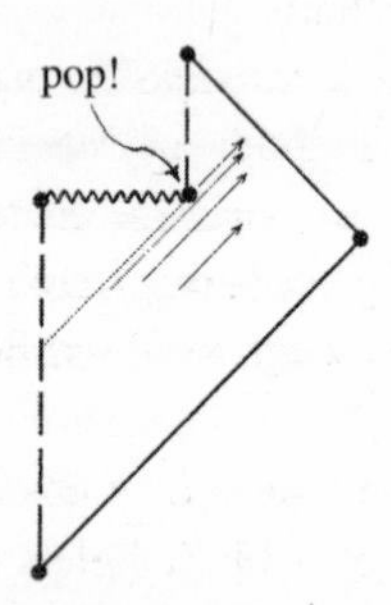

Fig. 2.41 Strict conformal diagram of Hawking-evaporating black hole.

Of course, most space-times do not possess spherical symmetry, and a description in terms of a strict conformal diagram may not even supply a reasonable approximation. Nevertheless, the notion of a *schematic* conformal diagram can frequently be of considerable value for clarifying ideas. Schematic conformal diagrams do not have the clear-cut rules that govern the strict ones, and sometimes one needs to imagine that the diagram is presented in 3 (or even 4) dimensions for its implications to be fully appreciated. The basic point is to make use of two of the ingredients of *conformal* representations of space-times which make infinite quantities finite. These are, on the one hand, the bringing into our finite comprehension the *infinite* regions of space and time that we have seen in our strict conformal diagrams, that have been depicted by bold-line boundaries and, on the other, the folding out of those regions that are infinite in a different sense, namely the *space-time singularities* that in our strict diagrams have been denoted by wiggly-line boundaries. The first has been achieved by a conformal factor (the 'Ω' of $\mathbf{g}\mapsto\Omega^2\mathbf{g}$ in §2.3) which has been allowed to tend smoothly to *zero*, so that the infinite regions are 'squashed' down to something finite. The second has been achieved, by a conformal factor that has been allowed to become *infinite*, so that the singular regions have been rendered finite and smooth by 'stretching out'. Of course, we may not be guaranteed that such procedures will actually work, in any particular case. Nevertheless, we shall find that both these procedures have important roles to play in the ideas that we shall be coming to, and the combination of the two will be central to what I am proposing in Part 3.

To end this section, it will be useful to present one context in which both these procedures can be particularly illuminating, namely with regard to the issue of *cosmological* horizons. In fact, there are two distinct notions that, in the cosmological context, are referred to as 'horizons'.[2.48] One of these is what is known as an *event* horizon; the other a *particle* horizon.

Let us first consider the notion of a cosmological event horizon. It is closely related to that of a black hole's event horizon, though the latter has a more 'absolute' character in the sense that it is less dependent upon some observer's perspective. Cosmological event horizons occur when the model possess a $\mathscr{I}^+$ which is *spacelike* as with all those Friedmann $\Lambda>0$ models exhibited in the strict conformal diagrams of Fig. 2.35 and in the de Sitter model $\mathbb{D}$ of Fig. 2.36(b), but the idea applies also in situations of a spacelike $\mathscr{I}^+$ where no symmetry is assumed (this being a general feature of $\Lambda>0$). In the schematic conformal diagrams of Fig. 2.42(a),(b), I have indicated (for 2 or 3 space-time dimensions, respectively) the region of space-time that is in principle observable to an observer O (considered to be immortal!) with world-line l terminating at a point o^+ on $\mathscr{I}^+$. This observer's *event horizon* $\mathcal{C}^-(o^+)$ is the past light cone of o^+.[2.49] Any event that occurs outside $\mathcal{C}^-(o^+)$ will forever remain unobservable to O. See Fig. 2.43. We notice, however, that the exact location of the event horizon is very much dependent on the particular terminal point o^+.

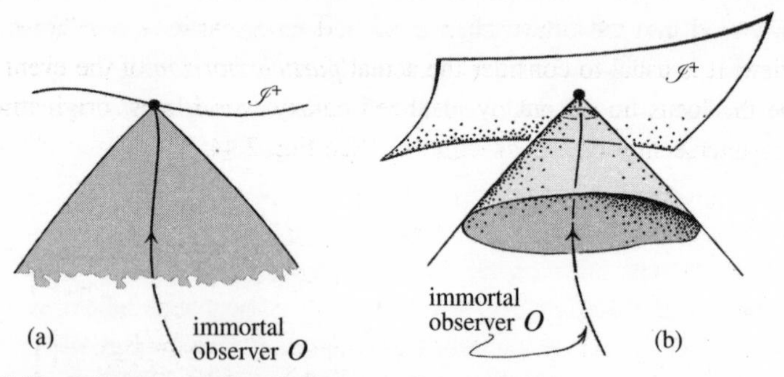

Fig. 2.42 Schematic conformal diagrams of cosmological event horizons occurring when $\Lambda > 0$: (a) 2-dimensional; (b) 3-dimensional.

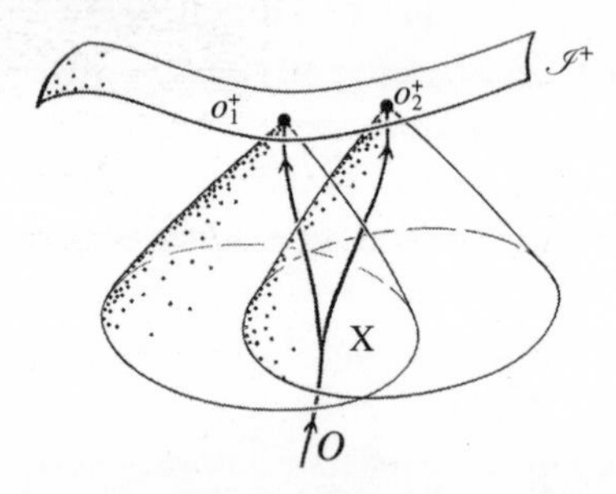

Fig. 2.43 The event horizon of the immortal observer O represents an absolute boundary to those events that are ever observable to O, this horizon itself dependent on O's choice of history. A change of mind at X can result in a different event horizon.

Particle horizons, on the other hand, arise when the *past* boundary—normally taken to be a singularity rather than infinity—is spacelike. In fact, as may be gleaned from those strict conformal diagrams depicted here in which singularities appear, a spacelike character is the norm for space-time singularities. This is closely related to the issue of 'strong cosmic censorship', which I shall touch upon in the next section. Let us call this initial singular boundary $\mathscr{B}^-$. If the event o is the space-time location of some observer O, then we may consider the past light cone $\mathscr{C}^-(o)$ of o, and see where it meets $\mathscr{B}^-$. Any particles that originate on $\mathscr{B}^-$ outside this intersection will never enter the region visible to the observer at the event o, although if O's world-line is allowed to be extended into the future, then more and more particles will come into view. It is usual to consider the actual *particle horizon* of the event o to be the locus traced out by idealized galaxy world-lines, originating at the intersection of $\mathscr{C}^-(o)$ with $\mathscr{B}^-$. See Fig. 2.44.

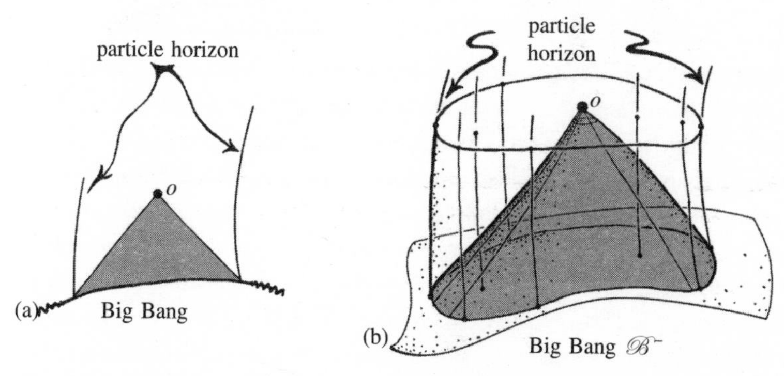

Fig. 2.44 Schematic conformal diagrams of particle horizons in (a) 2 dimensions, (b) 3 dimensions.

2.6 Understanding the way the Big Bang was special

Let us return to the basic question that we have been trying to address in this part, namely the issue of how our universe happened to come about with a Big Bang that was so extraordinarily special—yet special in what appears to have been a very peculiar way where, with regard to *gravity*, its entropy was enormously low in comparison with what it might have been, but the entropy was close to maximum in every other respect. This issue tends to be muddied, in most modern cosmological considerations, however, by the popular idea that in the very early stages of the universe's existence it underwent an exponential expansion—referred to as cosmic *inflation*—during a phase lasting for a tiny time-period somewhere between around 10^{-36} and 10^{-32} seconds following the Big Bang, increasing the linear dimension of the universe by some enormous factor of somewhere between 10^{20} and 10^{60}, or perhaps even 10^{100} or so. This huge expansion is supposed to explain the uniformity of the early universe (among other things), where practically all early irregularities are taken to have been ironed out simply by the expansion. However, these discussions seem hardly ever to be taken as addressing the fundamental question that I have been concerned with in Part 1, namely the origin of the extraordinary manifest specialness of the Big Bang, which *must* have been initially present, in order that there be a second law of thermodynamics. In my view, this idea underlying inflation—that the uniformity in the universe that we now observe should be the result of (inflationary) physical processes acting in its early evolution—is basically misconceived.

Why do I say that it is misconceived? Let us examine this issue in terms of some general considerations. The dynamics underlying inflation is taken to be governed in the same general way as are other physical processes, where there are *time-symmetrical* dynamical laws underlying this activity. There is taken to be a particular physical field known as the 'inflaton field' that is held to be responsible for inflation, although the precise nature of the equations governing the inflaton field would generally differ from one version of inflation to another. As part of the inflationary process, there would be some sort of 'phase transition' taking place, which may be thought of in terms of some kind of analogy with the transition between solid and liquid states that occurs with freezing or melting, etc. Such transitions would be regarded as proceeding in accordance with the Second Law, and would normally be accompanied by a raising of entropy. Accordingly, the inclusion of an inflaton field in the dynamics of the universe does not affect the essential arguments that were being put forward in Part 1. We still need to understand the extraordinarily low-entropy start of the universe, and according to the arguments of §2.2 this lowness of entropy lay essentially in the fact that the gravitational degrees of freedom were not excited, at least not nearly to the extent that involved all other degrees of freedom.

It will certainly be helpful to try to understand what a *high* entropy initial state would be like, when gravitational degrees of freedom are to be taken into consideration. We can get some appreciation of this if we imagine the time-reversed context of a *collapsing* universe, since this collapse, if taken in accordance with the Second Law, ought to lead us to a singular state of genuinely high entropy. It should be made clear that our mere *consideration* of a collapsing universe has nothing to do with whether our actual universe will ever re-collapse, like the closed $\Lambda=0$ Friedmann model of Fig. 2.2. This collapse is being taken simply as a hypothetical situation, and it is certainly in accord with Einstein's equations. In a general collapse situation, like the general collapses to a black hole that we considered in §2.4, we may expect all sorts of irregularities to emerge, but when local regions of material get sufficiently concentrated, trapped surfaces are likely to come about and space-time singularities are expected to arise.[2.50] Whatever density irregularities are present initially would intensify greatly, and the final singularity would be expected to be that coming from an extraordinary mess of congealing black holes. It is here that the considerations of Belinski, Khalatnikov and Lifshitz might well come into play. And if the BKL conjecture is correct (see §2.4), then some extremely complicated singularity structure is indeed to be expected.

I shall return to this issue of singularity structure shortly, but for the moment, let us consider the relevance of inflationary physics. Let us focus attention on the state of the universe at, for example, the time of decoupling, when the radiation that we now see as the CMB was produced (see §2.2). In our *actual* expanding universe, there was a very great uniformity in the matter distribution at that time. This is clearly taken to be a puzzle—for otherwise there would be no point in introducing inflation in order to explain it! Since it is accepted that there is something to explain, we must consider that there *might*, instead, have been enormous irregularities at that time. The inflationist's claim would have to be that the presence of an inflaton field actually renders such irregularities highly improbable. But is this really the case?

Not at all, for we can imagine this situation of a highly lumpy matter distribution at the time of decoupling, but with time reversed, so that this picture represents a very irregular *collapsing* universe.[2.51] As our imagined universe collapses inwards, the irregularities will become magnified, and deviations from FLRW symmetry (see §2.1) will become more and more exaggerated. Then, the situation will be so far from FLRW homogeneity and isotropy that the inflationary capabilities of the inflaton field will find no role, and (time-reversed) inflation will simply not take place, since this depends crucially on having an FLRW background (at least with regard to calculations that have actually been carried through).

We are therefore led to the clear implication that our irregular collapsing model will indeed collapse down to a state involving a horrendous mess of congealing black holes, this leading to a highly complicated enormously high-entropy singularity, very possibly of a BKL type, which is quite unlike the highly uniform low-entropy singularity of closely FLRW form that we seem to have had in our actual Big Bang. This would happen quite independently of whether or not an inflaton field is present in the allowed physical processes. Thus, time-reversing our imagined collapsing lumpy universe back again, so as to obtain a possible picture of an expanding universe, we find that it starts with a high-entropy singularity which, it seems, *could* have been an initial state for our actual universe and, indeed, would be a far more probable initial state (i.e. of much larger entropy) than the Big Bang that actually occurred. The black holes that congeal together in the final stages of our envisaged collapse would, when time-reversed to an expanding universe, provide us with the image of an initial singularity consisting of multiply bifurcating *white holes*![2.52] A white hole is the time-reverse of a black hole, and I have indicated the sort of situation that this provides us with in Fig. 2.45. It is the total *absence* of such white-hole singularities that singles out our Big Bang as being so extraordinarily special.

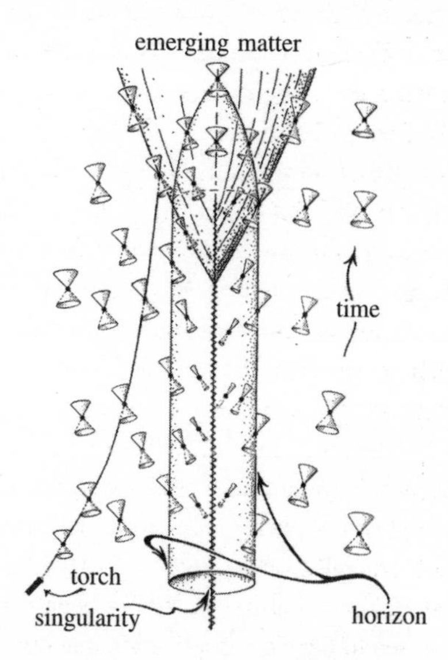

Fig. 2.45 A hypothetical 'white hole', which is the time-reverse of a black hole, such as depicted in Fig. 2.24. It violently disobeys the Second Law. Light cannot enter through the horizon, so light from the torch at the lower left gets in only after the hole explodes to ordinary matter.

In terms of phase-space volume, initial singularities of this nature (with multiply bifurcating white holes) would occupy a stupendously larger region than do those resembling the singularity that gave rise to our actual Big Bang. The mere potential presence of an inflaton field certainly cannot provide the power to 'iron out' the irregularities of such a conglomeration of white-hole singularities. This can be said with confidence quite apart from any detailed considerations of the nature of the inflaton field. It is just an issue of having equations that can be evolved equally in either direction in time, up until a singular state is reached.

But we can certainly say more about the actual enormity of the phase-space volume, if we take into account the entropy values, and therefore the phase-space volumes, that are actually assigned to black holes, according to the well-accepted Bekenstein–Hawking formula for the entropy value of a black hole. For a non-rotating hole of mass M, this entropy is

$$S_{BH}=\frac{8kG\pi^2}{ch}M^2,$$

whereas the entropy lies between this value and one-half of it if the hole is rotating, depending on the amount of the rotation. The fraction preceding 'M^2' is just a *constant*, where k, G, and h are the constants of Boltzmann, Newton, and Planck, respectively, c being the speed of light. In fact, we can rephrase this entropy formula in a more general form

$$S_{BH}=\frac{kc^3A}{4G\hbar}$$

where A is the surface area of the horizon and $\hbar=h/2\pi$, this formula being applicable whether or not the hole is rotating. In the Planck units to be introduced at the end of §3.2, we have

$$S_{BH}=A/4.$$

Although there is, in my opinion, still no completely satisfactory account of this entropy in terms of the counting of internal black-hole states,[2.53] such an entropy value is nevertheless an essential ingredient to the maintaining of a consistent Second Law in a quantum physical world external to the black hole. As mentioned already in §2.2, easily the largest contribution to the entropy of the present universe comes from the contributions from large black holes in galactic centres. If a total mass consisting of that lying within our present observable universe (that lying within our present particle horizon; see §2.5) were to form a black hole, this would attain an entropy of roughly 10^{124}, and we may consider this to provide a rough lower limit to entropy that would be achievable by our collapsing universe model involving the same amount of material. The phase-space volume corresponding to this would then be something like[2.54]

$$10^{10^{124}}$$

(because of the logarithm in Boltzmann's entropy formula, given in §1.3), whereas the region of phase space corresponding to the state of the actual observed universe at the time of decoupling, for the same body of matter, namely that in the observed CMB, had a volume no greater than about

$$10^{10^{89}}$$

The probability of finding ourselves in a universe of such a degree of specialness, if it had come about just by chance,[2.55] has the utterly absurdly tiny value of around $1/10^{10^{124}}$ irrespective of inflation. This is the kind of figure that needs some completely different kind of theoretical explanation!

There is, however, one further issue that may be considered to have importance here. This is the question of whether an initial singularity of such a complicated white-hole-type structure could reasonably be referred to as an 'instantaneous event'. The question is basically a matter of whether such a singularity, when viewed as some kind of past 'conformal boundary' to the space-time, can be appropriately thought of as 'spacelike'. Such a spacelike initial singularity could then be taken to represent the *zero* of some cosmic time coordinate and regarded as the 'moment' of such a highly irregular big bang.

In fact, the time-reverse of an Oppenheimer–Snyder collapse indeed has a spacelike initial singularity, as is clear from its strict conformal diagram Fig. 2.46, this being the time-reverse of Fig. 2.38(a). Moreover, it is a feature of general BKL singularities that they seem to have this spacelike character. More generally still, a spacelike nature is expected for generic singularities (allowing for their possibly being null in places) on the basis of *strong cosmic censorship*,[2.56] a yet unproved conjecture for solutions of Einstein's equations (referred to already in §2.4) which tells us that 'naked singularities' do not occur in generic gravitational collapse, the singularities that result being always hidden from direct observation, as by a black hole's event horizon. Strong cosmic censorship tells us that these singularities ought indeed to be spacelike, at least in general. In accordance with this expectation, it seems to me to be perfectly reasonable to refer to such a white-hole-ridden initial singularity as indeed being an instantaneous event.

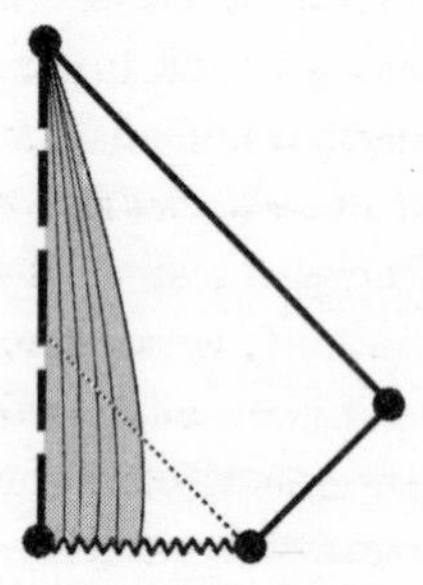

Fig. 2.46 Strict conformal diagram of the white hole in Fig. 2.45.

An important question now arises: what geometrical criterion distinguishes the kind of 'smooth' singularity that appears to be what characterized the very low-entropy singularity of our Big Bang, from the more general high-entropy type of singularity that arises in the white-hole-ridden time-reversed collapses just considered? We need some clear-cut way of saying that 'the gravitational degrees of freedom were not activated'. But for this, we need to identify the mathematical quantity that actually measures 'gravitational degrees of freedom'.

A good analogy for the gravitational field is the electromagnetic field, which resembles it in many significant ways, although there are, nevertheless, some important differences. The electromagnetic field is described, in relativity physics, by a tensor quantity **F**, referred to as the

Maxwell field tensor—after the great Scottish scientist James Clerk Maxwell, who first found, in 1861, the equations satisfied by the electromagnetic field, and he showed that these explain the propagation of light. We may recall that in §2.3, we encountered another tensor quantity, namely the *metric tensor* **g**. Tensors are essential for general relativity theory, as they provide mathematical descriptions of geometrical or physical entities in ways that are unaffected by (or 'carried along' by) the 'rubber-sheet' deformations (diffeomorphisms) that we considered in §2.3. The tensor **F** is determined by 6 independent numbers per point (3 for the components of the electric field at that point and 3 more for the magnetic field). The metric tensor **g** has 10 independent components per point. In standard tensor notation, it is usual to denote the collection of components of the metric by g_{ab}, or some such, with two lower indices (and it has a symmetry $g_{ab}=g_{ba}$). In the case of Maxwell's tensor **F**, the collection of components would be denoted by F_{ab} (with the *anti-symmetry* $F_{ab}=-F_{ba}$). Each of these tensors has a *valence* $[^0_2]$, which refers to the fact that there are just two lower indices. But tensors with upper indices can occur also, a $[^p_q]$-tensor being described by a collection of components denoted by an entity with p upper indices and q lower indices. There is an algebraic procedure known as *contraction* (or transvection) which allows us to connect a lower index to an upper one (rather in the manner of chemical bonding), thereby removing these two indices from the final expression—but it is not my purpose here to go into the algebraic operations of the tensor calculus.

The degrees of freedom in the electromagnetic field are indeed measured by the Maxwell tensor **F**, but in Maxwell theory there is also a *source* for the electromagnetic field, known as the *charge-current* vector **J**. This may be thought of as a $[^1_0]$-tensor, whose 4 components per point describe the 1 component of electric charge density together with the 3 components of electric current. In a stationary situation, the charge density acts as the source of electric field and the current density as the source for the magnetic field, but things get more complicated when the situation is not stationary.

We now ask for the analogues of **F** and **J** in the case of the *gravitational* field, as described by Einstein's general theory of relativity. In this theory there is a *curvature* to space-time (which can be calculated once one knows how the metric **g** varies throughout the space-time), described by a $\left[{0 \atop 4}\right]$-tensor **R**, called the *Riemann(–Christoffel)* tensor, with somewhat complicated symmetries resulting in **R** having 20 independent components per point. These components can be separated into two parts, constituting a $\left[{0 \atop 4}\right]$-tensor **C**, with 10 independent components, called the *Weyl conformal* tensor, and a symmetric $\left[{0 \atop 2}\right]$-tensor **E**, also with 10 independent components, called the *Einstein* tensor (this being equivalent to a slightly different $\left[{0 \atop 2}\right]$-tensor referred to as the *Ricci* tensor[2.57]). According to Einstein's field equations, it is **E** that provides the *source* to the gravitational field. This is normally expressed[2.58]in the form

$$\mathbf{E} = \frac{8\pi G}{c^4}\mathbf{T} + \Lambda\mathbf{g},$$

or, in the Planck units of §3.2, simply

$$\mathbf{E} = 8\pi\mathbf{T} + \Lambda\mathbf{g},$$

where Λ is the cosmological constant, and where the *energy* $\left[{0 \atop 2}\right]$-tensor **T** represents the mass-energy density and other quantities related to it via requirements of relativity; in other words, **E** (or equivalently, the energy tensor **T**) is the gravitational analogue of **J**. The Weyl tensor **C** is then the gravitational analogue of Maxwell's **F**.

We may ask what directly observable effects **C** and **E** might have, like magnetic fields being shown up by patterns of iron filings or by the pointing of a compass needle, and like electric fields being revealed by their effect on pith balls, etc. In fact, in an almost literal sense, we can actually *see* the effects of **E** and more particularly **C**, since these tensors have a direct and distinguishing effect on light rays—and in this respect **E** and **T** are *completely* equivalent, since $\Lambda\mathbf{g}$ has no effect on light rays. It may justly be said that the first clear evidence in support of general relativity was such a direct observation—which came from (Sir) Arthur Eddington's expedition to the Island of Principe in order to view, during the solar eclipse of 1919, the apparent displacement of stars' locations due to the Sun's gravitational field.

Basically, **E** acts as a magnifying lens, whereas **C** acts as a purely astigmatic lens. These effects are well described if we imagine how light rays are affected as they pass near or through a massive body, such as the Sun. Of course ordinary light will not actually propagate through the body of the Sun (or of the obscuring Moon, during the eclipse, for that matter), so we do not directly observe those particular rays in this case. But we can imagine that if we could actually see the star field through the Sun, then that field would be magnified slightly, owing to the presence of **E**, where the gravitating material of the Sun's actual body resides. The pure effect of **E** would be simply to *magnify* one's 'view' of what lies behind, without distortion.[2.59] However, when it comes to the distortion of the image of the distant star field *outside* the Sun's apparent disc (and this is what is *actually* observed) we find a gradual reduction of the outward displacement the farther out we look, and this leads to an astigmatic *distortion* of the distant star field. These effects are illustrated in Fig. 2.47. The distortion of the field outside the Sun's limb makes a small circular pattern in the distant star field appear elliptical, and this ellipticity is a measure of the amount of *Weyl curvature* **C** intercepted by the line of sight.

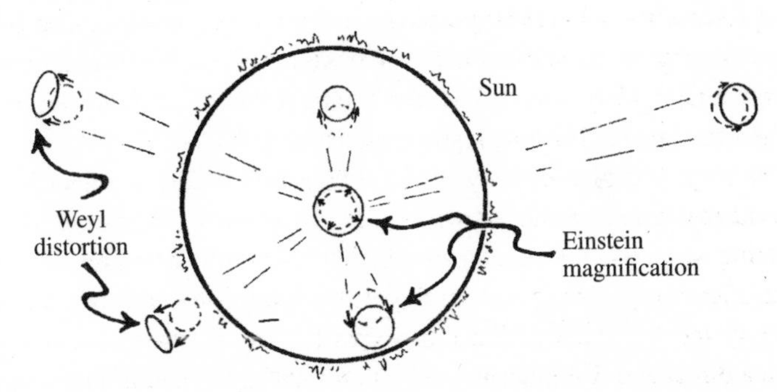

Fig. 2.47 The presence of Weyl curvature surrounding a gravitating body (here the Sun) can be seen in the distorting (non-conformal) effect that it has on the background field.

In fact, this gravitational lensing effect, originally predicted by Einstein, has become an extremely important tool in modern astronomy and cosmology, since it provides a means of measuring mass distributions that might even be otherwise completely invisible. In most of these cases, the distant background field consists of large numbers of very distant galaxies. The objective is to ascertain whether significant ellipticity has been introduced into the appearance of this background field, and to use this to estimate the actual intervening mass distribution whose gravitational field has caused the pattern of ellipticities. A snag, however, is that galaxies themselves tend to be rather elliptical, so one cannot usually tell whether or not an individual galaxy's image has been distorted. However, with large numbers of background field galaxies, statistics can be brought in, and often some very impressive estimates of mass distributions can be obtained in this way. On occasion, it is even possible to judge these things by eye, and some impressive examples are provided in Fig. 2.48, where the patterns of ellipticity make the presence of lensing sources particularly evident. One important application of this technique is in the mapping of *dark matter* distributions (see §2.1), since these are otherwise invisible.[2.60]

The fact that **C** introduces ellipticity into the images along light rays is indicative of its role as the quantity describing *conformal curvature*. At the end of §2.3 it was remarked that the conformal structure of space-time is in fact its null-cone structure. The conformal curvature of space-time, namely **C**, therefore measures the deviation of this null-cone structure from that of Minkowski space $\mathbb{M}$. We see that the nature of this deviation is that it introduces ellipticity into bundles of light rays.

Let us now come to the condition that we require, in order to characterize the very special nature of the Big Bang. Basically, we require a statement that gravitational degrees of freedom were unexcited at the Big Bang, which means saying something like 'the Weyl curvature **C** vanished there'. For many years, I have indeed been proposing that some such condition '**C**=0' holds at initial-type singularities, as opposed to what evidently happens in the 'final-type' singularities occurring in black holes for which **C** is likely to become infinite, as it does towards the singularity in the Oppenheimer–Snyder collapse, and perhaps diverging extremely wildly as in BKL singularities.[2.61] In general terms, this condition of the vanishing of **C** at initial-type

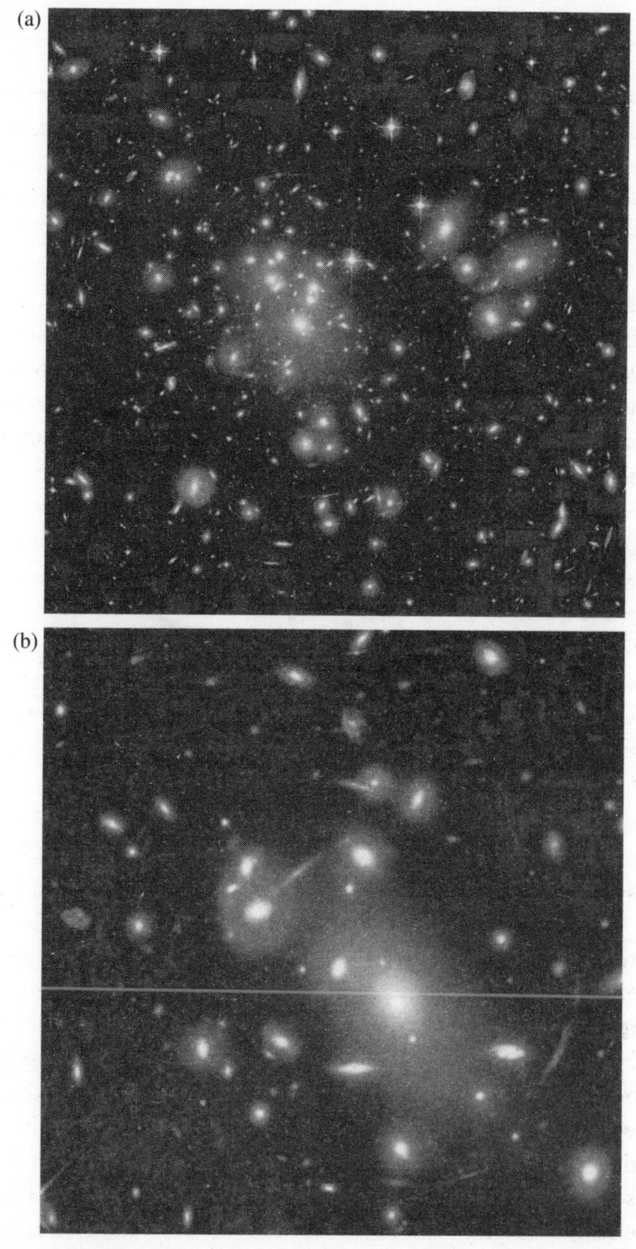

Fig. 2.48 Gravitational lensing: (a) galaxy cluster Abell 1689; (b) galaxy cluster Abell 2218.

singularities—which I have termed the *Weyl curvature hypothesis* (or WCH)—seems appropriate, but it is a little awkward that there are in fact numerous different versions of such a statement. The trouble is, basically, that **C** is a *tensor* quantity and it is hard to make unambiguous mathematical assertions about how such quantities behave at space-time singularities, where the very notion of a tensor, in any ordinary sense, loses its meaning.

It is fortunate, therefore, that my Oxford colleague Paul Tod has made a detailed study of a quite different, and mathematically much more satisfactory way of formulating a 'WCH'. This is to say, more or less, that there is a Big Bang 3-surface $\mathscr{B}^-$, which acts as a smooth past boundary to the space-time $\mathcal{M}$, when $\mathcal{M}$ is considered as a conformal manifold, just as happens in the exactly symmetrical FLRW models as is exhibited in the strict conformal diagrams of Fig. 2.34 and Fig. 2.35, but where the FLRW symmetry of these particular models is now *not* assumed. See Fig. 2.49. Tod's proposal at least constrains **C** to be *finite* at the Big Bang (since the conformal structure at $\mathscr{B}^-$, is assumed to be smooth), rather than **C** diverging wildly, and this statement might be taken to be sufficient for what is required.

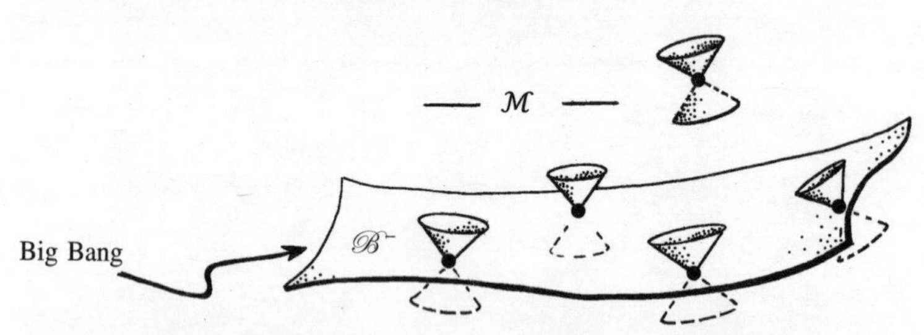

Fig. 2.49 Schematic conformal diagram of Paul Tod's proposal for a form of 'Weyl curvature hypothesis'; asserting that the Big Bang provides a smooth boundary $\mathcal{B}$ to the space-time $\mathcal{M}$.

To make this condition mathematically clearer, it is convenient to assert it in the form that the space-time can be continued smoothly, as a conformal manifold, a little way *prior* to the hypersurface $\mathscr{B}^-$. To *before* the Big Bang? Surely not: the Big Bang is supposed to represent the beginning of all things, so there can be no 'before'. Never fear—this is just a mathematical trick. The extension is not supposed to have any *physical* meaning!

Or might it . . .?

Part 3

Conformal cyclic cosmology

3.1 Connecting with infinity

Physically, what would the material universe have been actually like, far far back in time, very soon after the Big Bang? One thing in particular: it would have to have been *hot*—extremely hot. The kinetic energy in the motions of particles around at that time would have been so enormous as to have completely overwhelmed the particles' comparatively tiny rest energies ($E = mc^2$, for a particle of rest-mass m). Thus, the rest-mass of the particles would have been effectively irrelevant—as good as *zero* as far as the relevant dynamical processes are concerned. The contents of the universe, at extremely early times, would have consisted of effectively *massless* particles.

To phrase this issue in somewhat different terms, we may bear in mind that, according to current particle-physics ideas[3.1] about how the *masses* of basic particles actually come about, a particle's rest-mass ought to arise through the agency of a *special* particle (or perhaps a family of such special particles) referred to as the *Higgs* boson(s). Thus, the standard view about the origin of the rest-mass of any fundamental particle of Nature is that there is a quantum field associated with the Higgs that has the effect, through a subtle quantum-mechanical 'symmetry-breaking' procedure, of actually assigning a mass to other particles—a mass which they would not possess were it not for the Higgs. The Higgs would itself be thereby assigned its own particular mass (or, equivalently, rest energy). But in the very early universe, when the temperature was so high as to have provided energies greatly in excess of this Higgs value, *all* particles would then, according to standard ideas, indeed have become effectively massless, like a photon.

Massless particles, as we may recall from §2.3, do not appear to be particularly concerned with the full *metric* nature of space-time, respecting merely its *conformal* (or null-cone) structure. To be a little more explicit (and careful) about this, let us consider the primary massless particle—the photon—which, in fact, remains massless today.[3.2] To understand photons properly, we need to think of them in the context of the weird but precise theory of *quantum mechanics* (or, more correctly, quantum field theory, QFT). I cannot go into any details of QFT here (although I shall address some basic quantum issues in §3.4); our main concern is the physical *field*, of which photons provide the quantum constituents. This field is Maxwell's *electromagnetic* field, as described by the tensor **F**, referred to in §2.6. Now, it turns out that Maxwell's field equations are completely *conformally invariant*. What this means is that whenever we make the replacement of the metric **g** by a conformally related one $\hat{\mathbf{g}}$

$$\mathbf{g} \mapsto \hat{\mathbf{g}},$$

the new metric being (non-uniformly) rescaled

$$\hat{\mathbf{g}} = \Omega^2\, \mathbf{g},$$

where Ω is a positive-valued and smoothly varying scalar quantity on the space-time (see §2.3), we can find appropriate scaling factors for both the field **F** and its source, the charge-current vector **J**, so that exactly the same Maxwell equations hold as before,[3.3] but now with all operations defined in terms of $\hat{\mathbf{g}}$ rather than **g**. Accordingly, any solution of the Maxwell equations, with one particular choice of conformal scale, goes over to an exactly corresponding solution when any other choice of conformal scale is made. (This will be explained in slightly more detail in §3.2, and more fully in Appendix A6.) Moreover, at a primitive level, this is basically consistent with QFT,[3.4] in that the correspondence with the *particle* (i.e. photon) description also carries over to the hatted metric $\hat{\mathbf{g}}$, with individual photon going over to individual photon. Thus, the photon itself does not even 'notice' that a local scale change has been made.

Maxwell theory is, indeed, conformally invariant in this strong sense, where the electromagnetic *interactions* that couple electric charges with the electromagnetic field are also insensitive to local changes of scale. Photons, and their interactions with charged particles, do need space-time to have a *null-cone structure*—i.e. a conformal space-time structure—in order that their equations can be formulated, but they do *not* need the scale factor that distinguishes one actual metric from another, consistent with this given null-cone structure. Moreover, exactly the same invariance holds for the *Yang–Mills* equations that are considered to govern not only the *strong* interactions that describe the forces between nucleons (protons, neutrons, and their constituent quarks) and other relevant strongly interacting particles, but also the *weak* interactions that are responsible for radioactive decay. Mathematically, Yang–Mills theory[3.5] is basically just Maxwell theory with some 'extra internal indices' (see Appendix A7), so that the single photon is replaced by a multiplet of particles. In the case of strong interactions, things called *quarks* and *gluons* are the respective analogues of the electrons and photons of electromagnetic theory. The quarks, but not the gluons, are massive, with masses considered to be directly linked to the Higgs. In the standard theory of weak interactions (called 'electro-weak' theory, as electromagnetic theory is now also incorporated into this theory), the photon is considered to be part of a multiplet containing three other particles, all of which are massive, referred to as W^+, W^-, and Z. Again, these masses are considered to be coupled to that of the Higgs. Thus, according to current theory, when that mass-providing ingredient is removed, at the extremely high temperatures back near the Big Bang—and, indeed, roughly at the extremely high particle energies that are proposed would be reached by the LHC (Large Hadron Collider) particle accelerator in CERN, based in Geneva, when it is at full power[3.6]—then full conformal invariance should be restored. Of course, the details of this depend upon our standard theories of these interactions being appropriate, but this seems to be a not unreasonable assumption, as our ideas of particle physics stand at the moment. In any case, even if it turns out (for example when detailed results from the LHC become known and understood) that things are not quite as current theory suggests, it still remains probable that when energies get higher and higher, rest-masses become more and more irrelevant, physical processes becoming dominated by conformally invariant laws.

The upshot of all this is that close to the Big Bang, probably down to around 10^{-12} seconds after that moment,[3.7] when temperatures exceed about 10^{16} K, the relevant physics is believed to become blind to the scale-factor Ω, and *conformal* geometry becomes the space-time structure appropriate to the relevant physical processes.[3.8] Thus, all this physical activity would, at that stage, have been insensitive to local scale changes. In a conformal picture in which the Big Bang is stretched out, according to Tod's proposal of (§2.6, Fig. 2.49), to become a completely smooth spacelike 3-surface $\mathscr{B}^-$ which mathematically extends to a conformal 'space-time' *prior* to the Big Bang, the physical activity would propagate backwards in time in a mathematically coherent way, providing a physically sensible picture, seemingly unperturbed by the enormous scale changes involved, into this hypothetical pre-Big-Bang region that is being provided for it in accordance with Tod's proposal. See Fig. 3.1.

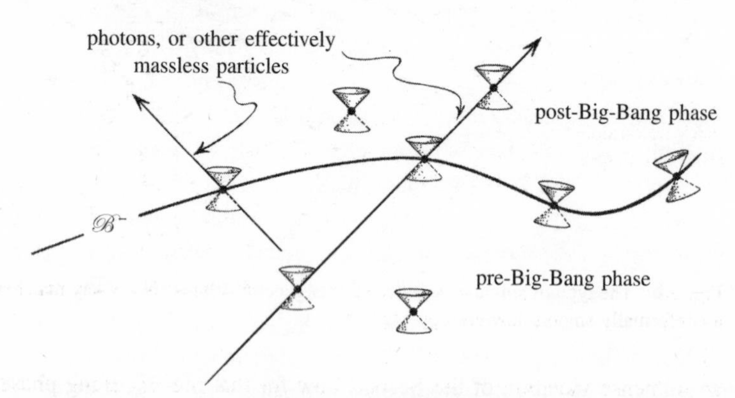

Fig. 3.1 Photons and other (effectively) massless particles/fields can propagate smoothly from an earlier pre-Big-Bang phase into the current post-Big-Bang phase or, conversely, we can propagate the particle/field information backwards from post- to pre-Big-Bang phase.

May we really suppose that we should be treating this hypothetical region as being actually physically *real*? If so, what kind of space-time region could this 'pre-Big-Bang' phase be? Perhaps the most immediate suggestion might be some collapsing phase of the universe which in some way is able to *bounce* back into an expanding universe at the Big Bang. But such a picture would negate all that I have been attempting to achieve up to this point. That picture would have our collapsing pre-Big-Bang phase somehow 'aimed' with incredible precision at such a very special ultimate state, of the same extraordinary degree of specialness that we appear to find in our actual Big Bang. It would represent

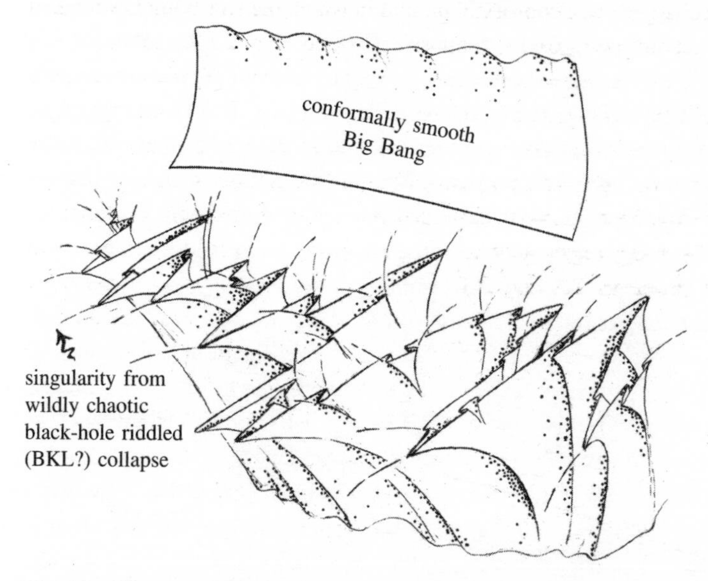

Fig. 3.2 The type of singularity expected in a generic collapse in no way matches a conformally smooth low-entropy big bang.

an immense violation of the Second Law for that pre-Big-Bang phase, with entropy reducing itself down to the (relatively) extremely tiny value that we find at the Big Bang. We recall the picture of a collapsing universe *in accordance* with the Second Law that was evoked in §2.6. This would be a thoroughly black-hole-riddled space-time that collapses to a singularity that in no way resembles a geometry with the required conformal smoothness needed for the kind of matching that Tod's proposal requires (see Fig. 3.2). Of course, one might adopt a viewpoint for which, in the pre-Big-Bang phase, the Second Law simply operates the other way around in time (cf. the final paragraphs of §1.6), but that goes very much against the grain of the overall purpose behind the enterprise undertaken by this book. The hope is to find something more like an 'explanation' of the Second Law, or at least some kind of rationale for it, rather than simply *decreeing* that some absurdly special state occurs at some stage during the universe's history (namely at the 'bounce' moment being considered above). Moreover, it turns out that there are also some *mathematical* difficulties with this particular kind of a 'bounce' proposal, as we shall be seeing later (in §3.3, in relation to Tolman's radiation-filled universe models; see also Appendix B6).

No, let us try something very different. Let us try to examine the other end of time, namely what is expected in the extremely remote future. According to the models described in §2.1 in which there is a positive cosmological constant Λ (see Fig. 2.5), our universe ought ultimately to settle into an exponential expansion, apparently rather closely modelled by the strict conformal diagrams of Fig. 2.35, in which there is a smooth spacelike future conformal boundary $\mathscr{I}^+$. Of course, our own universe now possesses certain types of irregularity, the greatest *local* departures from the highly symmetrical FLRW geometry being the presence of black holes, especially the very massive ones at galactic centres. However, in accordance with the discussion of §2.5, all black holes ought eventually to disappear with 'pops' (see Fig. 2.40 and its strict conformal diagram Fig. 2.41), even though the very largest holes might have to take something like a googol (i.e. ~ 10^{100}) or more years before this happens.

Following that extremely long time-span, the physical contents of the universe will, in terms of numbers of particles, consist mainly of photons, these coming from greatly red-shifted starlight and CMB radiation, and from the Hawking radiation that will ultimately carry away almost the entire mass-energy of numerous huge black holes, in the form of very low-energy photons. But there will also be gravitons (the quantum constituents of gravitational waves) coming from close encounters between such black holes, especially the very big holes in galactic centres—and these encounters will actually turn out to play a vital role for us in §3.6. Photons are massless particles, but so also are gravitons, and neither of these can be used to make a clock, in accordance with the disscussion of §2.3, as illustrated in Fig. 2.21.

There will presumably also be a good measure of 'dark matter' around, whatever that mysterious substance might be (§2.1, and see also §3.2 for my own general proposal), to the extent that this material would have survived capture by black holes. It is hard to see how such a substance, interacting only through the gravitational field, could be of much value in the construction of a clock. To take such a standpoint would, however, represent a subtle change of philosophy; yet, we shall be seeing in §3.2, that such a subtle change will in any case be a necessary feature of the overall picture that I shall be presenting. Thus, it again begins to seem that it might be just the *conformal* structure of space-time that would, in the ultimate stages of our universe's expansion, be what is physically relevant.

When the universe enters this apparently final stage—what one might well call the 'very boring era'—nothing of great interest seems to be left for it to do. The most exciting events prior to this were the final 'pops' of the last tiny remnants of black holes, eventually disappearing (it is supposed) after they had very gradually lost all their mass via the painfully slow process of Hawking radiation. One is left with the dreadful thought of a seemingly interminable boredom confronting the final stages of our great universe—a universe which would have once seemed so exciting, teeming with fascinating activity of hugely different kinds—most of this activity occurring within beautiful galaxies, with a wonderful variety of stars and often attendant planets, among which would be those supporting life of some kind, with its exotic plants and animals, some of whom having the capabilities of deep knowledge and understanding, and profound capabilities of artistic creation. Yet all this will eventually die away. The final dregs of excitement will have to be the waiting, and the waiting, and waiting, for maybe 10^{100} years or more, for that final pop—perhaps of about the violence of a small artillery shell followed by nothing but further exponential expansion, thinning it out and cooling and emptying and cooling, and thinning out . . . until eternity. Does that picture present all that our universe has ultimately in store for it?

But after I had been depressing myself with such thoughts, one day in the summer of 2005, another thought then occurred to me, which was to ask: who will be around then to be bored by this apparent overpowering eventual tedium? Surely not us; it will be mainly massless particles like photons and gravitons. And it is pretty hard to bore a photon or a graviton—even aside from the extreme unlikelihood that such entities could actually *have* significant experiences! The point is that, according to a massless particle, the passage of time is as nothing. Such a particle can even *reach* eternity (that is, $\mathscr{I}^+$) before encountering the first 'tick' of its internal clock, as was illustrated in Fig. 2.22. One might well say that 'eternity is no big deal' for a massless particle such as a photon or a graviton!

To put this another way, it would appear that *rest-mass* is a necessary ingredient for the building of a clock, so if eventually there is little around which has any rest-mass, the capacity for making measurements of the passage of time would be lost (as is the capacity for making distance measurements, since distances also depend on time measurements; see §2.3). Indeed, as we have seen before, massless particles do not appear to be particularly concerned with the *metric* nature of space-time, respecting merely its *conformal* (or null-cone) structure. Accordingly, to massless particles, the ultimate hypersurface $\mathscr{I}^+$ represents a region of their conformal space-time that seems to be just like anywhere else, and there appears to be no bar to their entering a hypothetical extension of this conformal space-time on the 'other side' of $\mathscr{I}^+$. Moreover, there are powerful mathematical results, mainly through the important work of Helmut Friedrich,[3.9] that lend support to the actual conformal future-extendability of space-time, under the general circumstances being considered here, for which there must be a positive cosmological constant Λ.

This mirrors our discussion of the physics at a Big-Bang hypersurface which accords with Tod's proposal. It appears that (for different reasons) *both* $\mathscr{I}^+$ *and* $\mathscr{B}^-$ would be likely to allow smooth extensions of the conformal space-time to regions on the other sides of these hypersurfaces. Not only that, but the material contents on either side would be likely to be an essentially *massless* substance whose physical behaviour is basically governed by conformally invariant equations, and this would enable the activity of this material to be continued into both of these hypothetical extensions of (conformal) space-time.

One possibility might indeed suggest itself at this point. Could it be that our $\mathscr{I}^+$ and $\mathscr{B}^-$ are one and the same? Perhaps, as a conformal manifold, our universe just 'loops round', so that what lies beyond $\mathscr{I}^+$ is simply our own universe starting up again from its Big-Bang origin, conformally stretched out as $\mathscr{B}^-$, according to Tod's proposal. The economy of this idea certainly has its appeal, but I think that there could be serious difficulties of consistency which, in my own view, render this suggestion implausible. Basically, such a space-time would contain *closed timelike curves* whereby causal influences can lead to potential paradoxes, or at least to unpleasant constraints on behaviour. Such paradoxes or constraints do depend upon the possibility of coherent information being able to pass across the $\mathscr{I}^+/\mathscr{B}^-$ hypersurface. Yet we shall be seeing in §3.6 that this kind of thing is a real possibility in the type of scheme that I am proposing here, and so such closed timelike curves do indeed have the potential to lead to serious inconsistency problems.[3.10] For reasons such as this I am *not* proposing this $\mathscr{I}^+/\mathscr{B}^-$ identification.

However, I am suggesting 'the next best thing', which is to propose that there *is* a physically real region of space-time prior to $\mathscr{B}^-$ which is the remote future of some previous universe phase, and that there is *also* a physically real universe phase that extends beyond our $\mathscr{I}^+$ to become a big bang for a new universe phase. In accordance with this proposal, I shall refer to the phase beginning with our $\mathscr{B}^-$ and extending to our $\mathscr{I}^+$ as the present *aeon*, and I am suggesting that the universe as a whole is to be seen as an extended conformal manifold consisting of a (possibly infinite) succession of aeons, each appearing to be an entire expanding universe history. See Fig. 3.3. The '$\mathscr{I}^+$' of each is to be identified with the '$\mathscr{B}^-$' of the next, where the continuation of each aeon to the next is achieved so that, as a *conformal* space-time structure, the join is perfectly smooth.

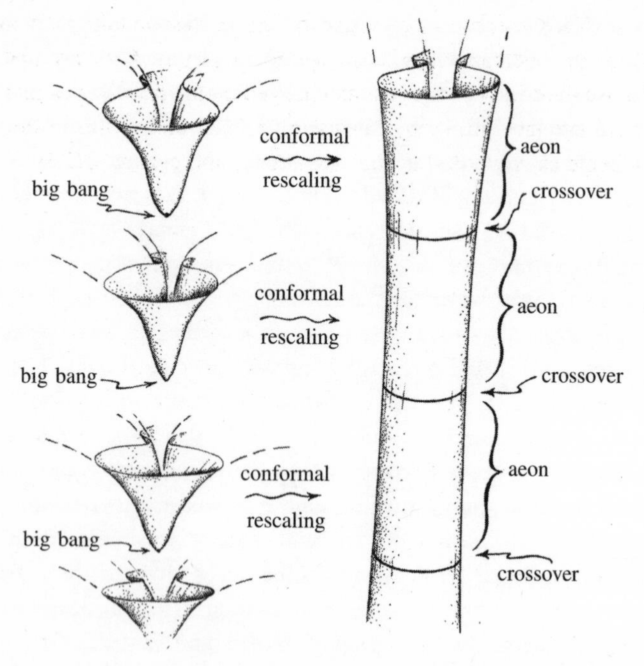

Fig. 3.3 Conformal cyclic cosmology. (As with my drawing in Fig 2.5, I am trying not to prejudice the issue of whether the universe is spatially open or closed.)

The reader might well worry about identifying a remote future, where the radiation cools down to zero temperature and expands out to zero density, with a big-bang-type of explosion, where the radiation had started at an infinite temperature and infinite density. But the conformal 'stretching' at the big bang brings this infinite density and temperature down to finite values, and the conformal 'squashing' at infinity brings the zero density and temperature up, to finite values. These are just the kinds of rescalings that make it possible for the two to match, and the stretching and squashing are procedures that the relevant physics on either side is completely insensitive to. It may also be mentioned that the *phase space* $\mathcal{P}$, describing the totality of possible states of all the physical activity on either side of the crossover (see §1.3), has a *volume measure* which is conformally invariant,[3.11] basically for the reason that when distance measures are reduced, the corresponding momentum measures are increased (and *vice versa*) in just such a way that the product of the two is completely unchanged by the rescaling (a fact that will have crucial significance for us in §3.4). I refer to this cosmological scheme as *conformal cyclic cosmology*, abbreviated CCC.[3.12]

3.2 The structure of CCC

There are various aspects of this proposal that require a good deal more detailed attention than I have given above. One key issue concerns what the *full* contents of the universe might be likely to be in the very remote future. The discussion above concentrated mainly on the considerable background of photons that would be present, from starlight, from the CMB, and from black-hole Hawking evaporation. I have also considered that there would be a significant contribution to this background from gravitons, by which I mean the basic (quantum) constituents of gravitational waves, these waves being 'ripples' of space-time curvature, arising largely from close encounters between extremely large black holes in galactic centres.

Photons and gravitons are both massless, so it seems not unreasonable to adopt a philosophy, relevant to the very remote future, that since, in a very late stage in the universe's history it would in principle be impossible to build a clock out of such material, then the universe itself, in the remote future, would somehow 'lose track of the scale of time' and so the geometry of the physical universe really becomes *conformal geometry* (i.e. null-cone geometry), rather than the full metric geometry of Einstein's general relativity. In fact, we shall be seeing shortly that there are subtleties in connection with the gravitational field which compel us to moderate this philosophy somewhat. But for the moment, let us confront another difficulty with this philosophical standpoint which needs to be faced.

When considering what the main contents of the universe might be in the late stages of its existence, I have ignored the fact that there would be much material within bodies that do not ever find themselves within a black hole, having been flung out from their parent galaxies through random processes, where in some cases the body would also escape from the galactic cluster within which it had originally resided and where there would, indeed, also be much dark matter that would never fall into a black hole. What, for example, would be the fate of a white dwarf star that had escaped in this way, cooled down to become an invisible black dwarf? It has often been suggested that protons might eventually decay away, though observational limits tell us that the rate at which this could happen would have to be very slow indeed.[3.13] In any case, there would be decay products of some kind, and although much of the material of the black dwarf might eventually collapse into a black hole via such processes, there would be likely to be many 'rogue' massive particles that had, in some form, escaped from the clusters of galaxies to which they had originally been attached.

My concern is particularly with *electrons*—and also with their anti-particles, the *positrons*—because they are the least massive *electrically charged* particles. It is not a particularly unconventional view that protons, and other charged particles more massive than electrons and positrons, might eventually, after vast periods of time, decay into less massive particles. We might imagine that all protons could ultimately decay in this way, but if we accept the conventional view that electric charge must be absolutely conserved, then the ultimate decay products of a proton must contain a net positive charge, so that at least one positron would be expected to be among the eventual survivors. A similar argument would apply to negatively charged particles, and it is hard to escape the conclusion that there would have to be numerous electrons present as well, to accompany these positrons. There might also be more massive charged particles such as protons and anti-protons, if these do *not* eventually decay, but the key problem lies with the electrons and positrons.

Why is this a problem? Could there not be another type of charged particle (one both of a positive and of a negative charge) which is actually *massless*, so that electrons and positrons could eventually decay into these, and the above philosophical standpoint be retained? The answer appears to be 'no'. For the mere existence of such a type of massless charged particle, among the menagerie of particle types taking part in today's physical activities, would have made its presence copiously manifest in numerous particle processes.[3.14] Yet, these processes are actually seen to take place *without* the production of such massless charged particles. Consequently, there are *no* massless charged particles around today. Will the (massive) electrons and positrons then have to be around until eternity, in contradiction with the intended philosophical standpoint?

One possibility for retaining this standpoint is raised by the thought that the remaining electrons and positrons might seek each other out and eventually mutually annihilate one another completely to produce merely photons, which would then be harmless to this philosophy. But, unfortunately, in the extremely remote future, many individual charged particles will find themselves isolated within their cosmological event horizons, as shown in Fig. 3.4 (see also Fig. 2.43 in §2.5), and when that happens—as it sometimes must—it removes any possibility of such an eventual charge annihilation. A possible resolution would be to weaken our philosophical standpoint somewhat, and to argue that the odd electron or positron, trapped within its event horizon, would hardly be of much use for the construction of an actual clock. For my own part, I am dissatisfied with such a line of reasoning, as it seems to me to lack the kind of rigour that physical laws ought to demand.

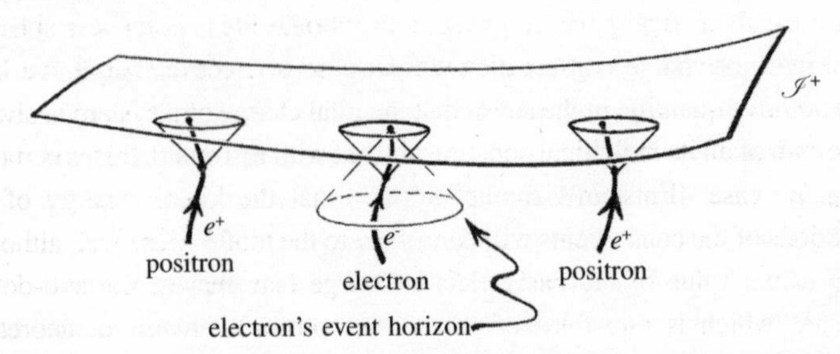

Fig. 3.4 There will be the occasional 'rogue' electron or positron, ultimately trapped within its horizon and unable to lose its electric charge through pair annihilation.

A more radical resolution might be to suppose that charge conservation is actually *not* one of Nature's stringent requirements. Accordingly, it might be the case that, at extremely occasional moments, a charged particle might decay into one that is without electric charge, and over the reaches of eternity, all electric charge could, accordingly, eventually vanish away. On this consideration, electrons or positrons might eventually become converted into one of their uncharged siblings, say a *neutrino*, in which case it would also be a requirement that, among the three known types of neutrino, there is one without rest-mass.[3.15] Quite apart from there being no evidence whatever for any violation of charge conservation, such a possibility is an extremely unpleasant one, theoretically, and it would also seem to demand that the photon itself acquire a small mass, which would in itself nullify the proposed philosophical standpoint.

The one remaining possibility that occurs to me, and which actually strikes me as something to be considered seriously, not merely the least of all evils, is that the notion of rest-mass is not the absolute constant that we imagine it to be. The idea is that over the reaches of eternity, the surviving massive particles—the electrons, positrons, neutrinos, and also protons and antiprotons, if they do not eventually decay, and moreover whatever might be the constituent of the dark matter (necessarily without charge, but possessing rest-mass)—would find that their very rest-masses would very, very gradually fade away, attaining the value zero in the eventual limit. Again, there is absolutely no observational evidence, as of now, for such a violation of ordinary notions concerning rest-mass, but in this case the theoretical backing of the conventional ideas is far less substantial than for charge conservation. In the case of electric charge, we have an *additive* quantity, in the sense that the total charge of a system is always the sum of all its individual constituents, but with rest-mass, this is certainly not the case. (Einstein's $E=mc^2$ tells us that the kinetic energy of the motions of the constituents will contribute to the total.) Moreover, although the actual value of the basic electric charge (say that of the anti-down-quark, which is one third of that of the proton) remains a theoretical mystery, the values of all other charges found in the universe are whole-number multiples of this value. Nothing like this appears to be the case for rest-mass, and the underlying reason for the particular values of the rest-masses of individual particle types is completely unknown. So there appears to be still the freedom that the rest-mass of a fundamental particle is not an absolute constant—as indeed it is not, according to standard particle physics, in the very *early* universe, as remarked above, in §3.1—and that it might indeed fade away to zero in the very remote future.

In relation to this, one final technical comment may be made concerning the status of rest-mass in particle physics. A standard procedure for addressing the idea of an 'elementary particle' is to look for what are termed the 'irreducible representations of the Poincaré group'. Any elementary particle is supposed to be described according to such an irreducible representation. The *Poincaré group* is the mathematical structure describing the symmetries of Minkowski space $\mathbb{M}$, and this procedure is a natural one in the context of special relativity and quantum mechanics. The Poincaré group possesses two quantities referred to as *Casimir operators*,[3.16] these being *rest-mass* and *intrinsic spin*, and accordingly the rest-mass and spin are deemed to be 'good quantum numbers', which remain constant so long as the particle is a stable one and does not interact with anything. However, this role of $\mathbb{M}$ appears to be less fundamental when there is a positive cosmological constant Λ present in physical laws (as $\Lambda=0$ for $\mathbb{M}$), and it would seem that, when we are concerned with matters related to cosmology, it should be the symmetry group of *de Sitter space-time* $\mathbb{D}$, rather than of $\mathbb{M}$, that should ultimately be our concern (see §2.5, Fig. 2.36(a),(b)). However, it turns out that rest-mass is *not* exactly a Casimir operator of the de Sitter group (there being a small additional term involving Λ), so that its ultimate status is more questionable in this case, and a very slow decay of rest-mass seems to me to be not out of the question.[3.17]

The extremely gradual decaying away of rest-mass, according to this proposal, does have its curious implications, however, with regard to the whole scheme of CCC, because it raises a new issue in relation to the measurement of time. We recall that near the end of §2.3 a particle's rest-mass was used to provide a well-defined scale of time, such a scaling being all that is needed so that we may pass from a conformal structure to a full metric. If, as seems to be required from the above discussion, we need particles' masses to decay away, albeit extremely gradually, then we are led into a bit of a quandary. Do we still adopt this idea of using particles' rest-masses for precisely defining our space-time's metric, when massive particles are still around, but with slowly decaying masses? If we try to settle on some particular particle type, say an electron, as providing us with the standard of time, then with the kind of decay rates that would seem to be required in order for electrons to be considered adequately 'massless' when $\mathscr{I}^+$ is reached (see Appendix A2), it would turn out that $\mathscr{I}^+$ is *not* at infinity at all, and the universe's expansion, according to this 'electron metric' would either have to slow to a halt or else to reverse into a collapse. It would appear that such behaviour would not be consistent with Einstein's equations. Moreover, if instead of an 'electron metric' we used a 'neutrino metric' or 'proton metric', say, then the detailed geometrical behaviour of the space-time would be likely to differ from the corresponding behaviour that would be obtained by use of electrons (unless the scaling to zero occurs with all mass values retaining exactly their initial proportions). To me, this does not appear very satisfactory.

It seems that in order to preserve some appropriate form of Einstein's equations—with constant Λ—throughout the entire history of the aeon, we need to use another proposal for scaling for the metric. What we can do, although this would hardly be a 'practical' solution for the purposes of building a clock, would be to use Λ itself to determine a scale, or, what appears to be closely related to this, we might use the effective value of the gravitational constant G. Then the picture of an evolving and unendingly exponentially expanding universe continuing into its remote future would be retained, but without seriously disturbing the philosophy that, locally, the universe will eventually lose track of the scale of time.

This matter is closely related to another one which I have glossed over until now, namely the fact that whereas there is a conformal invariance for the free gravitational field, as described by the Weyl conformal tensor **C** (since **C** indeed describes the conformal curvature), the coupling of the field to its sources is *not* conformally invariant. This is quite different from what happens in Maxwell's theory, where there is a conformal invariance which holds *both* for the free electromagnetic field **F** and for the coupling between **F** and its sources as described by the charge-current vector **J**. Thus, again, when we bring *gravity* into the picture in a serious way, the basic philosophy of CCC gets a little muddied. We must take the view that, in a sense, the philosophy of CCC asserts that it is *gravity-free* physics (and Λ-free physics) that loses track of time, not completely physics as a whole.

Let us try to understand the relation of Einstein's theory to conformal invariance. It is a somewhat delicate matter. In the case of electromagnetism, the entire equations are preserved under the conformal rescaling. We are to examine what happens when the space-time metric **g** is replaced by a conformally related one $\hat{\mathbf{g}}$ by means of a *scale factor* Ω, this being a positive number varying smoothly over space-time (see §2.3, §3.1):

$$\mathbf{g} \mapsto \hat{\mathbf{g}} = \Omega^2\,\mathbf{g}.$$

To see the conformal invariance of Maxwell theory, we adopt rescalings for the $[{}^{0}_{2}]$-tensor **F** describing the field, and for the $[{}^{1}_{0}]$-tensor **J** describing the (charge-current) source, given by

$$\mathbf{F} \mapsto \hat{\mathbf{F}} = \mathbf{F} \text{ and } \mathbf{J} \mapsto \hat{\mathbf{J}} = \Omega^{-4}\,\mathbf{J}.$$

Maxwell's equations can be written symbolically as

$$\nabla\mathbf{F} = 4\pi\,\mathbf{J},$$

where ∇ stands for a specific set of differential operators[3.18] determined by the metric **g**. When the scale change $\mathbf{g} \mapsto \hat{\mathbf{g}}$ is applied, ∇ must be replaced by an operator quantity $\hat{\nabla}$, determined correspondingly by $\hat{\mathbf{g}}$, and then what we find (Appendix A6) is

$$\hat{\nabla}\hat{\mathbf{F}} = 4\pi\,\hat{\mathbf{J}}$$

which, being just the same equation as before but now in 'hatted' form, expresses the *conformal invariance* of Maxwell's equations. In particular, when $\mathbf{J} = 0$ all we have are the *free* Maxwell equations:

$$\nabla\mathbf{F} = 0,$$

and when $\mathbf{g} \mapsto \hat{\mathbf{g}}$ is applied, we find conformal invariance in

$$\hat{\nabla}\hat{\mathbf{F}} = 0.$$

This (conformally invariant) set of equations governs the propagation of *electromagnetic waves* (light) and it can also be regarded as the quantum-mechanical Schrödinger equation satisfied by individual free photons (see §3.4 and Appendix A2, A6).

In the case of gravity, the source $\left[{0 \atop 2}\right]$-tensor **E** (Einstein tensor, taking the place of **J**; see §2.6) does not have a scaling behaviour which provides conformal invariance for the equations, but there *is* a conformally invariant analogue of $\nabla\mathbf{F}=0$, which governs the propagation of gravitational waves, and provides an analogous Schrödinger equation for individual free gravitons. I shall write this symbolically (see Appendix A2, A5, A9) as

$$\nabla\mathbf{K}=0,$$

the subtlety here being that whereas this $\left[{0 \atop 4}\right]$-tensor **K** is taken to be identical to the Weyl conformal $\left[{0 \atop 4}\right]$-tensor **C** (of §2.6)

$$\mathbf{K}=\mathbf{C}$$

when the original (Einstein) physical metric **g** is used, we find (Appendix A9) that when we rescale to a new metric $\hat{\mathbf{g}}$ according to $\mathbf{g}\mapsto\hat{\mathbf{g}}=\Omega^2\mathbf{g}$, we must adopt different scalings

$$\mathbf{C}\mapsto\hat{\mathbf{C}}=\Omega^2\,\mathbf{C} \quad \text{and} \quad \mathbf{K}\mapsto\hat{\mathbf{K}}=\Omega\mathbf{K},$$

in order to preserve the *meaning* of **C** as providing the measure of conformal curvature, and to preserve the conformal invariance of the wave propagation of **K**, so that we get

$$\hat{\nabla}\hat{\mathbf{K}}=0,$$

then these scalings lead us to[3.19]

$$\hat{\mathbf{K}}=\Omega^{-1}\hat{\mathbf{C}}.$$

This has some curious consequences, which are of considerable importance for CCC. As we approach $\mathscr{I}^+$ from its past, we need to use a conformal factor Ω which tends to zero smoothly,[3.20] but with a non-zero normal derivative. The geometrical meaning of this is illustrated in Fig. 3.5. The conformal invariance of the wave-propagation equation for **K** implies that it attains *finite* (and usually non-zero) values on $\mathscr{I}^+$, these values determining the strength (and polarization) of the *gravitational radiation*—the gravitational analogue of light—as it continues out to infinity and thereby makes its mark on $\mathscr{I}^+$. See Fig. 3.6. The same applies to the values of **F** on $\mathscr{I}^+$, determining the strength and polarization of the electromagnetic radiation field (light). But because of the fact that Ω becomes zero at $\mathscr{I}^+$, the displayed equation above, rewritten as $\hat{\mathbf{C}}=\Omega\hat{\mathbf{K}}$, tells us that the finiteness of $\hat{\mathbf{K}}$ implies that the conformal tensor $\hat{\mathbf{C}}$ must itself become *zero* on $\mathscr{I}^+$ (where we use a metric $\hat{\mathbf{g}}$ finite at $\mathscr{I}^+$). Since $\hat{\mathbf{C}}$ provides a direct measure of conformal geometry at $\mathscr{I}^+$, the demand of CCC that the conformal geometry be *smooth* over the crossover 3-surface from each aeon to the next tells us that conformal curvature must also be zero at the *big-bang* surface $\mathscr{B}^-$ of the subsequent aeon. Accordingly, CCC actually provides a *stronger* version of Weyl curvature hypothesis (WCH, see §2.6) than the condition that the conformal curvature merely be finite (which was what Tod's proposal gave directly), namely that this conformal curvature really does *vanish* at the $\mathscr{B}^-$ of each aeon, in accordance with the original idea of WCH.

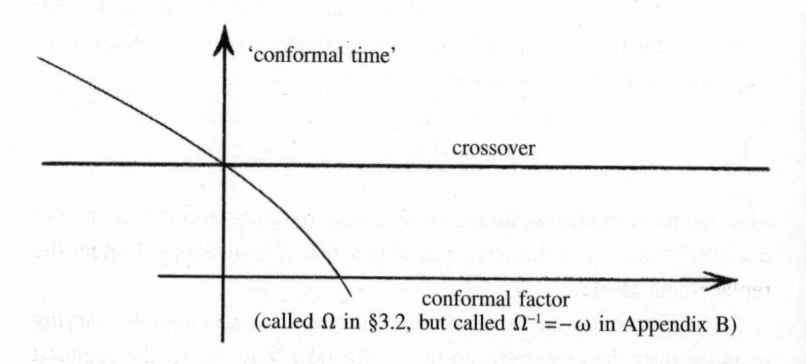

Fig. 3.5 The conformal scale factor goes cleanly from positive to negative at crossover, the curve having a slope that is neither horizontal nor vertical. Here 'conformal time' just refers to 'height' in a suitable conformal diagram.

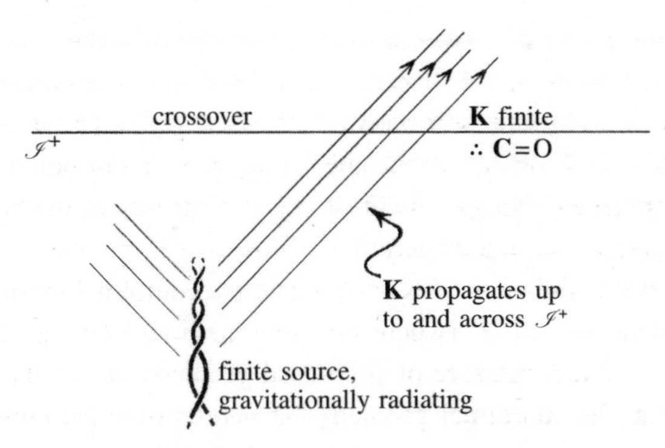

Fig. 3.6 The gravitational field is measured by the tensor **K**, propagates according to a conformally invariant equation, and so generally attains finite non-zero values at $\mathscr{I}^+$

On the other side of the crossover surface, i.e. just following the $\mathscr{B}^-$ of the subsequent aeon, we find a conformal factor which becomes *infinite* at $\mathscr{B}^-$, but in just such a way as to make Ω^{-1} behave smoothly at $\mathscr{B}^-$.[3.21] Thus, it appears to be the case that Ω has to be able to be continued somehow over the crossover 3-surface to become, suddenly, its *reciprocal*! A way to handle this situation mathematically is to encode the essential information of Ω in a way that does not distinguish it from its reciprocal Ω^{-1}. This can be done by considering the $[^0_1]$-tensor $\mathbf{\Pi}$ (a 1-form), that mathematicians would write as[3.22]

$$\mathbf{\Pi} = \frac{d\Omega}{\Omega^2 - 1}$$

The two most important things about $\mathbf{\Pi}$ are, first, that it remains smooth over the crossover 3-surface and, second, that it is unchanged under the replacement $\Omega \mapsto \Omega^{-1}$.

In CCC, we try to demand that $\mathbf{\Pi}$ indeed be a smoothly varying quantity over the crossover, so that if we take $\mathbf{\Pi}$ to define the required scaling information, rather than Ω, then we can imagine that the transition $\Omega \mapsto \Omega^{-1}$ at crossover can be achieved while $\mathbf{\Pi}$ remains smooth across it. This requires certain mathematical conditions to be satisfied for the behaviour of Ω at $\mathscr{I}^+$, and the indications are that these can indeed be achieved satisfactorily and uniquely. (Detailed arguments are given in Appendix B.) The upshot of it all is that there turns out to be a clear-cut and apparently unique mathematical procedure for continuing the massless fields into the future through the crossover 3-surface, it being assumed that only massless fields are present in the very remote future of the earlier aeon (i.e. just prior to $\mathscr{I}^+$).

With only massless fields present, we have a particular scaling freedom in the choice of rescaled metric $\hat{\mathbf{g}}$ in the region just prior to the $\mathscr{I}^+$ of the earlier aeon, consistent with its given conformal structure. This freedom is described in terms of a field ϖ, which satisfies a self-coupled (i.e. non-linear) conformally invariant massless scalar field equation that I refer to (in Appendix B2) as the 'ϖ-equation'. The different solutions of the ϖ-equation give us the different possible metric scalings that would get us from our chosen $\hat{\mathbf{g}}$-metric to the other possible metrics $\varpi^2\hat{\mathbf{g}}$ which Einstein's equations (with cosmological constant Λ) would tell us refer only to sources which are massless. The particular choice of ϖ that gives us Einstein's original physical metric $\mathbf{g}$, is referred to as the 'phantom field' (since in Einstein's $\mathbf{g}$-metric it disappears, simply taking the value 1). The phantom field does not have any independent physical degrees of freedom, in the region prior to the $\mathscr{I}^+$, but just keeps track of the metric $\mathbf{g}$, telling us the scaling that gets us back to $\mathbf{g}$ from the $\hat{\mathbf{g}}$-metric that is currently being used.

On the opposite side of the crossover, immediately following the big bang of the subsequent aeon, we find that simply continuing the fields smoothly through leads to an effective gravitational constant in this new aeon that has become negative, with unphysical implications. Consequently it becomes necessary to adopt the alternative interpretation, in which we use the alternative choice Ω^{-1}, consistent with Π, on the other side. This has the effect of turning the phantom field ϖ into a *real* physical field (albeit initially infinite) on the big-bang side of crossover. It is tempting to interpret this ϖ-field following this big bang as providing the initial form of new *dark matter*, prior to it acquiring a mass. Why make such an interpretation? The reason simply is that the mathematics forces there to be *some* dominant new contribution, of the nature of a scalar field, in the big bang of the new aeon, this arising from the above behaviour of the conformal factor. This is additional to the contributions from photons (electromagnetic field) or from any other particles of matter (considered to have lost their rest-mass by the time that they reach the crossover 3-surface). It has to be there for mathematical consistency, as soon as we adopt the $\Omega \mapsto \Omega^{-1}$, transformation at crossover.

An additional feature that we find coming out of the mathematics is that on the big-bang side of crossover, the condition that all sources are massless cannot be strictly maintained, although a natural constraint restricting unwanted freedom in the conformal factor is that this appearance of rest-mass is put off for as long as possible. Thus, a component to the post-big-bang matter content is this contribution bearing rest-mass. It would be natural to assume that this has something to do with the Higgs field (or whatever might turn out to be necessary) in its role in the appearance of rest-mass in the early universe.

Dark matter is the dominant form of matter, apparently observed to be present in the initial stages of our own aeon. It comprises some 70% of ordinary matter (where 'ordinary' just means *not* counting the contribution of the cosmological constant Λ—commonly referred to as 'dark energy'[3.23]), but dark matter does not seem to fit at all comfortably into the standard model of particle physics, its interaction with other kinds of matter being solely through its gravitational effect. The phantom field ϖ in the late stages of the prior aeon arises as an effective scalar component to the gravitational field, coming about only because we are allowing the conformal rescalings $\mathbf{g} \mapsto \Omega^2\mathbf{g}$, and it has no independent degrees of freedom. In the subsequent aeon, the new ϖ-matter that comes about initially takes over the degrees of freedom present in the gravitational waves in the prior aeon. Dark matter seems to have had a special status at the time of our Big Bang, and this is certainly the case for ϖ. The idea is that shortly after the Big Bang (presumably when the Higgs comes into play), this new ϖ-field acquires a mass, and it then becomes the actual dark matter that appears to play such an important role in shaping the subsequent matter distributions, with various kinds of irregularities that are observed today.

It is perhaps significant that the two so-called 'dark' quantities ('dark matter' and 'dark energy'), that have gradually become apparent from detailed cosmological observations in recent decades, both appear to be necessary ingredients of CCC. This scheme would certainly *not* work without $\Lambda > 0$, since the consequent *spacelike* nature of $\mathscr{I}^+$ is needed in order to match the spacelike character of $\mathscr{B}^-$. Moreover, we see from the above that the scheme requires that there be some sort of initial matter distribution which might reasonably be identified with the dark matter. It will be interesting to see whether this interpretation of dark matter will hold up theoretically and observationally.

With regard to Λ, the main thing that appears to puzzle cosmologists and quantum field theorists is its *value*. The quantity $\Lambda\mathbf{g}$ is often interpreted by quantum field theorists as the *energy of the vacuum* (see §3.5). For reasons to do with relativity, it is argued that this 'vacuum energy' ought to be a $\left[{0 \atop 2}\right]$-tensor proportional to $\mathbf{g}$, but the proportionality factor comes out as something larger than the observed value of Λ by a factor of around 10^{120}, so something is clearly missing from this idea![3.24] Another thing that is found puzzling is that Λ's observed tiny value is just such as to be starting to have effects on the expansion of the universe that are comparable with the particular totality of attraction due to matter in the universe *now*, which was enormously greater in the past and which will become enormously smaller in the future, and this seems to be an odd coincidence.

To me this 'coincidence' is not such an enormous puzzle, at least over and above some puzzles that had already been with us long before the observational evidence indicating Λ's actual small value. Certainly the observed value of Λ needs explanation, but perhaps it can be specifically related to the gravitational constant G, the speed of light c and Planck's constant h by some fairly simple formula, but with the 6th power of a certain large number N in the denominator

$$\Lambda \approx \frac{c^3}{N^6 G \hbar}.$$

Here

$$\hbar = \frac{h}{2\pi}$$

is Dirac's form of Planck's constant h (sometimes called the *reduced* Planck constant). The number N is about 10^{20} and it was pointed out, in 1937, by the great quantum physicist Paul Dirac that various integer powers of this number seem to turn up (approximately) in several different ratios of basic physical dimensionless constants, particularly when gravity is in some way involved. (For example, the ratio of the electric to the gravitational force between the electron and the proton in a hydrogen atom is around $10^{40} \approx N^2$.) Dirac also pointed out that the age of the universe is about N^3, in terms of the absolute unit of time that is referred to as the *Planck time* t_P. The Planck time, and the corresponding *Planck length* $l_P = ct_P$, are often regarded as providing a kind of 'minimum' space-time measure (or 'quantum' of time and space, respectively), according to common ideas about quantum gravity:

$$t_P = \sqrt{\frac{G\hbar}{c^5}} \approx 5.4 \times 10^{-44}\ \text{s}, \quad l_P = \sqrt{\frac{G\hbar}{c^3}} \approx 1.6 \times 10^{-35}\ \text{m}.$$

By use of these 'Planck units', and also the *Planck mass* m_P and Planck energy E_P given by

$$m_P = \sqrt{\frac{\hbar c}{G}} \approx 2.1 \times 10^{-5}\ \text{g}, \quad E_P = \sqrt{\frac{\hbar c^5}{G}} \approx 2.0 \times 10^{9}\ \text{J},$$

which are naturally determined (though completely impractical) units, one can express many other basic constants of Nature simply as pure (dimensionless) numbers. In particular, in these units, we have $\Lambda \approx N^{-6}$.

In addition, we can use Planck units for *temperature*, by setting Boltzmann's constant $k = 1$, where one unit of temperature is the absurdly large 2.5×10^{32} K. When considering the very large entropies involved with large black holes or with regard to the universe as a whole (as in §3.4), I shall use Planck units. However, for values this large, it turns out to make little difference what units are used.

Originally, Dirac thought that since the age of the universe is (obviously) increasing with time, then N ought to be increasing with time or, equivalently, G reducing with time (in proportion to the reciprocal of the square of the universe's age). However, more accurate measurements of G than were available when Dirac put his ideas forward have shown that G (or equivalently N), *if* it is not constant, cannot vary at the rate that Dirac's ideas required.[3.25] However, in 1961, Robert Dicke (with a refined later argument by Brandon Carter[3.26]) pointed out that according to the accepted theory of stellar evolution, the lifetime of an ordinary 'main-sequence' star is related to the various constants of Nature in such a way that any creature whose life and evolution depends upon its being around somewhere roughly in the middle of the time-span of such an ordinary star's active existence, would be likely to find a universe whose age, in Planck time units, is indeed around N^3. So long as the particular N^{-6} value of Λ can be theoretically understood, this would also explain the puzzle of the apparent coincidence of a cosmological constant coming into play just around now. Yet, these are clearly speculative matters, and admittedly some better theories will be required to provide understanding of these numbers.

3.3 Earlier pre-Big-Bang proposals

The scheme of CCC may be contrasted with a number of other proposals for pre-Big-Bang activity, which had been put forward previously. Even among the earliest cosmological models consistent with Einstein's general relativity, namely those of Friedmann put forward in 1922, there was one that became referred to as the 'oscillating universe'. This terminology seems to have arisen from the fact that for the closed Friedmann model without cosmological constant ($K>0$, $\Lambda=0$; see Fig. 2.2(a)), the *radius* of the 3-sphere that describes the spatial universe, when represented as a function of time, has a graph that has the shape of a *cycloid*, which is the curve traced out by a point on the circumference of a circular hoop rolling along the time axis (normalized so that the speed of light $c=1$ (see Fig. 3.7). Clearly this curve extends beyond the single arch that would describe a spatially closed universe expanding from its big bang and then collapsing back to its big crunch, where now we have a succession of such things, and we can think of the entire model as representing an unending succession of 'aeons' (Fig. 3.8), a scheme which briefly interested Einstein in 1930.[3.27] Of course the 'bounce' that takes place at each stage at which the spatial radius reaches zero occurs at a *space-time singularity* (where the space-time curvature becomes infinite) and Einstein's equations cannot be used in the ordinary way to describe a sensible evolution, even though some sort of modification might be envisaged, perhaps along lines something like those of §3.2.

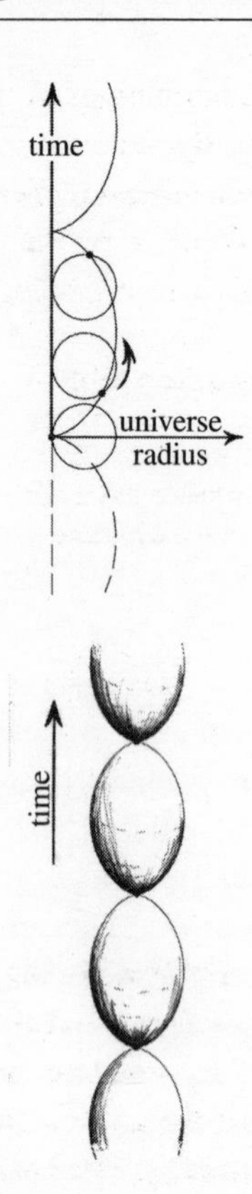

Fig. 3.7 The Friedmann model of Fig. 2.2(a) has a radius which, when plotted as a function of time, describes a cycloid, which is the curve traced out by a point on a rolling hoop.

Fig. 3.8 Taking the cycloid of Fig 3.7 seriously, we obtain an oscillating closed universe model.

A more serious matter, however, from the point of view of this book, is how such a model can address the issue of the Second Law, since this particular model leaves no scope for a progressive change representing a continual increase of entropy. In fact, in 1934, the distinguished American physicist Richard Chace Tolman described a modification of Friedmann's oscillating model[3.28] which alters Friedmann's 'dust' to a composite gravitating material with an additional internal degree of freedom, which can undergo changes to accommodate an increase in entropy. Tolman's model somewhat resembles the oscillating Friedmann model, but the successive aeons have progressively longer durations and increasingly greater maximum radii (see Fig. 3.9). This model is still of the FLRW type (see §2.1), so there is no scope for contributions to the entropy through gravitational clumping. Consequently, the entropy increase in this model is of a comparatively very mild kind. Nevertheless, Tolman's contribution was important in being one of the surprisingly few serious attempts to accommodate the Second Law into cosmology.

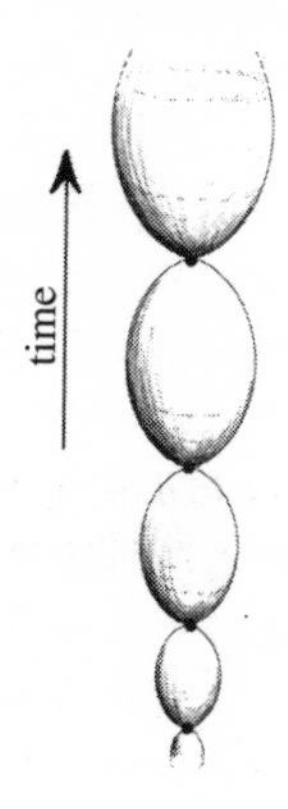

Fig. 3.9 This model, due to Tolman, began to address the Second Law by having a matter content allowing an increasing entropy, whence the model would become larger at each stage.

At this point, it is appropriate to mention another of Tolman's contributions to cosmology, which has some considerable relevance also to CCC. The representation of the material contents of the universe as a pressureless fluid (i.e. 'dust'; see §3.1) is the way that the gravitational *source* (i.e. the Einstein tensor **E**; see §2.6) is dealt with in the Friedmann models. This is not such a bad first approximation, so long as the actual material being modelled is reasonably dispersed and *cold*. But, when we are considering the situation in the close vicinity of the Big Bang, we need to treat the material contents as being very *hot* (see beginning of §3.1), so it is expected that a much better approximation near the Big Bang is *incoherent radiation*—although for the evolution of the universe following the time of decoupling (§2.2), Friedmann's *dust* is better. Accordingly, Tolman introduced radiation-filled analogues of all six of the Friedmann models of §2.1 in order to provide a better description of the universe close to the Big Bang. The general appearance of the Tolman radiation solutions does not differ greatly from that of the corresponding Friedmann solutions, and the pictures of Figs. 2.2 and 2.5 will do well enough also for the Tolman radiation solutions. The strict conformal diagrams of Fig. 2.34 and Fig. 2.35 will also do for the respective Tolman radiation solutions, except that the picture Fig. 2.34(a) needs to be replaced, strictly speaking, by one where the depicted rectangle is replaced by a *square*. (In the drawing of strict conformal diagrams, there is frequently enough freedom to accommodate such differences in scale, but in this case, the situation turns out to have a little too much rigidity to remove the global scale difference between these two figures.)

The cycloidal arch of Friedmann's Fig. 3.7, for the case $K>0$, must, in Tolman's radiation model, be replaced by the *semicircle* of Fig. 3.10, describing the universe's radius as a function of time ($K>0$). It is curious that the natural (analytic) continuation of Tolman's semicircle behaves quite differently from what happens with the cycloid, since a semicircle ought really to be completed to a *circle* if we are thinking of a genuine analytic continuation.[3.29] This makes no sense if we are trying to think of an actual continuation to values of the time-parameter which extend beyond the range of the original model. Basically, the universe radius would have to become *imaginary*[3.30] in Tolman's case, if we were to try to extend it analytically to a phase prior to this model's big bang. Thus a direct analytical continuation to provide a 'bounce' of the type that occurs with the 'oscillating' Friedmann $K>0$ solution does not seem to make sense when we move from Friedmanns's dust to Tolman's radiation, the latter being much more realistic for behaviour near the actual Big Bang, owing to the extraordinarily high temperature that we expect to find there.

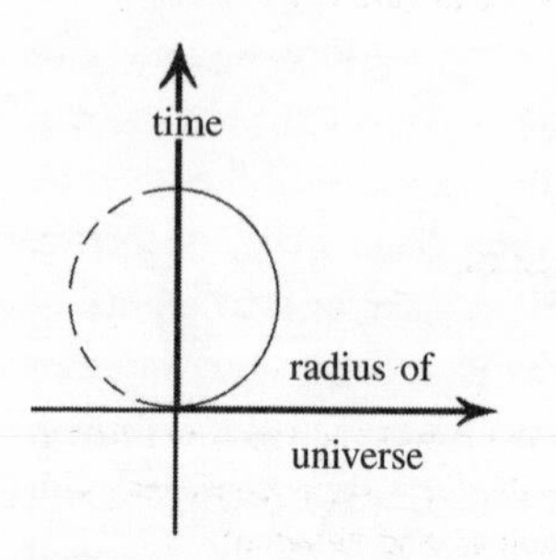

Fig. 3.10 In Tolman's closed radiation-filled universe, the radial function is a semicircle.

This difference in behaviour at the singularity has importance in relation to Tod's proposal (§2.6). This has to do with the nature of the conformal factor Ω that is needed to 'blow up' the big bangs of the Friedman solutions and corresponding Tolman radiation solutions to a smooth 3-surface $\mathscr{B}$. Since such an Ω becomes *infinite* at $\mathscr{B}$, it will be clearer if we phrase things in terms of the *reciprocal* of this Ω, for which I shall use the small letter ω:

$$\omega = \Omega^{-1}.$$

(The reader may be reassured, here, that despite the confusion between the notation used here and in Appendix B concerning the definition(s) of Ω, the ω used here actually *agrees* with that of Appendix B.) In the Friedmann cases, we find that close to the 3-surface $\mathscr{B}$, the quantity ω behaves like the *square* of a local (conformal) time parameter (vanishing at $\mathscr{B}$), so the continuation of ω across $\mathscr{B}$ is achieved smoothly without ω changing sign. Hence its inverse Ω does not become negative across $\mathscr{B}$ either; see Fig. 3.11(a). On the other hand, in the Tolman radiation cases, ω varies in *proportion* to such a local time parameter (vanishing at $\mathscr{B}$), so smoothness in ω would require the sign of ω, and therefore of Ω itself, to change to a *negative* value on one or the other side of $\mathscr{B}$. In fact, this latter behaviour is much closer to what happens in CCC. We saw, in §3.2, that a smooth conformal continuation of the remote expanded

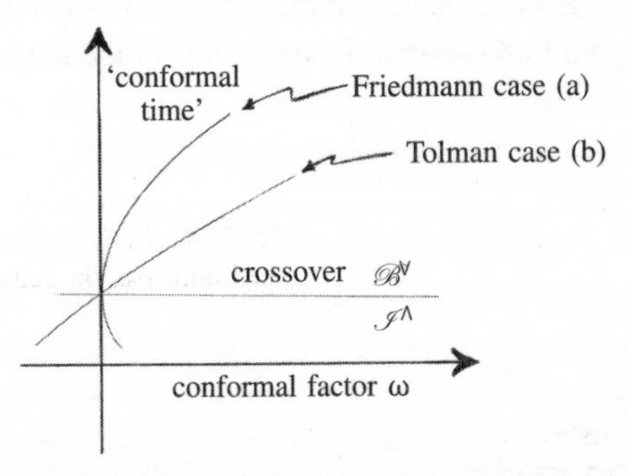

Fig. 3.11 Comparison between the behaviours of the conformal factors ω for (a) Friedmann's dust and (b) Tolman's radiation. Only the latter (b) is consistent with CCC. (See Fig. 3.5 and Appendix B for the terminology and notation.)

future of the aeon previous to the crossover 3-surface continues across having a negative Ω-value in the subsequent aeon (Fig. 3.11(b)). This gives us a catastrophic reversal of the sign of the gravitational constant, if we do not make the switch $\Omega \mapsto \Omega^{-1}$ at the crossover surface (see §3.2). But if we do make this switch, then the behaviour of $(-)\Omega$ on the big-bang side of crossover is necessarily the kind of behaviour that we get for a Tolman radiation solution, rather than a Friedmann one. This appears to be very satisfactory, because a Tolman radiation model indeed provides a good local approximation to the space-time immediately following the Big Bang (where I am ignoring the possibility of inflation, for reasons referred to in §2.6, §3.4 and §3.6).

There is a further idea that some cosmologists have proposed might be incorporated into cyclic models such as the Friedmann oscillating model of Fig. 3.8 or some modification of it like that due to Tolman illustrated in Fig. 3.9. This idea appears to have been originated by John A. Wheeler, when he put forward the intriguing proposal that the dimensionless constants of Nature might become altered when the universe passes through a singular state like the zero-radius moments that occur in these oscillating type models. Of course, since the normal dynamical laws of physics have had to be abandoned in order to get the universe through these singular states, there seems no reason why we should not abandon a few more and let the basic constants vary too!

But there is a serious point here. It has been frequently argued that there are many curious coincidences in the relations between the constants of Nature upon which life on Earth seems to depend. Some of these might be readily dismissed as being of value only to certain kinds of life we are familiar with, like the parameters determining the delicate fact that as ice forms from water, it is anomalous in being less dense than the water, so that life can persist in water remaining unfrozen under a protective surface layer of ice even when the external temperature drops below freezing. Others seem to present a more problematic challenge, such as the threat that the whole of chemistry would have been impossible had not the neutron been just marginally more massive than the proton, a fact which leads to a whole variety of different kinds of stable nuclei—these underlying all the different chemical elements—that would not otherwise have come about. One of the most striking of such apparent coincidences was revealed with William Fowler's confirmation of Fred Hoyle's remarkable prediction of the existence of a particular energy level of carbon which, had it not existed, would have meant that the production of heavy elements in stars would not have been able to proceed beyond carbon, leaving the planets devoid of nitrogen, oxygen, chlorine, sodium, sulphur, and numerous other elements. (Fowler shared in the 1982 Nobel Prize for this with Chandrasekhar but, strangely, Hoyle was passed over.)

The term 'anthropic principle' was coined by Brandon Carter, who made a serious study of the notion[3.31] that had the constants been not exactly right in this particular universe, or in this particular place or particular time in this particular universe, then we would have had to have found ourselves in another, where these constants did have suitable values for intelligent life to be possible. It is not my intention to pursue this extremely intriguing but highly contentious set of ideas further here. I am not altogether sure what my own position on the matter is, though I do believe that too much reliance is frequently placed on this principle in attempts to give support to what are, to me, implausible-sounding proposed theories.[3.32] Here, I merely point out that in passing from one aeon to the next in accordance with CCC, there might well be scope for changes to, say, the value of the 'N' referred to in §3.2, whose powers seem to determine the various ratios between widely differing fundamental dimensionless physical constants. The matter will be addressed again in §3.6.

Wheeler's idea is also incorporated into a more exotic proposal put forward by Lee Smolin in his 1997 book *Life of the Cosmos*.[3.33] Smolin makes the tantalizing suggestion that when black holes form, their internal collapsing regions—through unknown quantum-gravity effects—become converted to expanding ones by some kind of 'bounce', each one providing the seed of a new expanding universe phase. Each new 'baby universe' then expands to a 'full-grown' one with its own black holes, etc., etc. See Fig. 3.12. This collapse⇝expansion procedure would clearly have to be quite unlike the kind of conformally smooth transition involved in CCC (see Fig. 3.2), and its relation to the Second Law is obscure. Nevertheless the model has the virtue that it can be studied from the point of view of the biological principle of *natural selection*, and it is not entirely without significant statistical predictions. Smolin makes worthy attempts at such predictions and provides comparisons with observational statistics of black holes and neutron stars. The role of Wheeler's idea here is that the dimensionless constants might change only moderately in each collapse⇝expansion process, so that there would be some kind of 'inheritance' in the propensity to form new black holes, this being subject to the influences of a kind of natural selection.

Fig. 3.12 Smolin's romantic view of the universe where new 'aeons' emerge from black-hole singularities.

Hardly less fanciful, in my own humble view, are those cosmological proposals which depend for their operation on the ideas of *string theory* and their dependence—as string theory stands—on the existence of extra space dimensions. The earliest such pre-big-bang proposal, as far as I am aware, is one due to Gabriele Veneziano.[3.34] This model does seem to have some strong points in common with CCC (pre-dating CCC by some seven years), particularly in relation to the roles of conformal rescalings, and the idea that the 'inflationary period' might better be thought of as an exponential expansion occurring in a universe phase prior to the one we are presently experiencing (see §3.4, §3.6). On the other hand, it is dependent on ideas from string-theory culture, which makes it hard to relate directly to the CCC proposal being put forward here, particularly in relation to the clear-cut predictive elements of CCC that I shall come to in §3.6.

Similar remarks apply also to the more recent proposal of Paul Steinhardt and Neil Turok,[3.35] in which the transition from one 'aeon' to the next takes place via the 'collision of D-branes', D-branes being structures within a higher-dimensional adjunct to the normal notion of 4-dimensional space-time. Here, the crossover is taken to occur only at some smallish multiple of 10^{12} years, when all black holes that are currently believed to come about through astrophysical processes would still be around. Moreover, apart from this, the dependence on concepts from the string-theory culture again make it difficult to make clear-cut comparisons with CCC. This could be greatly clarified if their scheme could be reformulated in such a way that it can be viewed as being based on a more conventional 4-dimensional space-time, the roles of the extra-dimensional structures being somehow codified into 4-dimensional dynamics, even if only approximately.

In addition to the schemes referred to above, there are numerous attempts to use ideas from *quantum gravity* to achieve a 'bounce' from a previously collapsing universe phase to a subsequent expanding one.[3.36] In these, it is taken that a non-singular quantum evolution replaces the singular state that *classically* would occur at the moment of minimum size. In many attempts to achieve this, simplified lower-dimensional models are often used, although the implications for 4-dimensional space-time are then not altogether clear. Moreover, in most attempts at a quantum evolution, the singularities are still *not* removed. The most successful proposal to date for a non-singular quantum bounce appears to be that using the *loop-variable* approach to quantum gravity, and a quantum evolution through what would classically be a cosmological singularity has been achieved on these terms, by Ashtekar and Bojowald.[3.37]

However, as far as I am able to tell, none of the pre-Big-Bang proposals described in this section makes any serious inroad into the fundamental issue raised by the Second Law, as described in Part 1. None explicitly addresses the question of suppressing gravitational degrees of freedom in the Big Bang, this actually being the *key* to the origin of the Second Law in the particular form that we find it, as was emphasized in §2.2, §2.4 and §2.6. Indeed, most of the above proposals lie firmly within the scope of FLRW models, so they do not come close to addressing these essential matters.

Yet, even the early twentieth-century cosmologists were certainly aware that things might get very different as soon as deviations from FLRW symmetry are allowed. Einstein himself had expressed the hope[3.38] that the introduction of *irregularities* might enable the singularity to be avoided (rather in the same spirit of the much later work of Lifshitz and Khalatnikov, before they and Belinski located their error; see §2.4). As has now become clear, following the singularity theorems of the later 1960s,[3.39] this hope cannot be realized within the framework of classical general relativity, and models of this type will inevitably encounter space-time singularities. We see, moreover, that when such irregularities in the collapsing phase are present, and are growing relentlessly in accordance with the vast entropy increases that accompany gravitational collapse according to the picture presented in §2.6, then there is no possibility that the geometry—even just the conformal (null-cone) geometry—that the collapsing phase will attain at its big crunch can match the far smoother (FLRW-like) big bang of the subsequent aeon.

Accordingly, if we are to take the view that the pre-Big-Bang phase would indeed have to behave in accordance with the Second Law, with gravitational degrees of freedom becoming fully activated, then it would appear that something very different from a straightforward bounce, either classical or quantum, must be involved. My own attempt to address this serious question constitutes one of the principal reasons for putting forward the apparently somewhat strange idea of CCC—involving, as it does, the infinite scale change that permits the required geometrical matching between one aeon and the next. Yet, a profound puzzle still remains: how can such a cyclic process be nevertheless consistent with the Second Law, with entropy continually increasing throughout aeon after aeon after aeon ...? This challenge is central to the entire undertaking of this book and I shall need to confront it seriously in the next section.

3.4 Squaring the Second Law

Let us therefore return to the question which started out this whole enterprise, namely the origin of the Second Law. The first point to be made is that there is a conundrum to be faced. It is a conundrum that appears to confront us *irrespective* of CCC. The issue has to do with the evident fact that the entropy of our universe—or the current aeon, if we are considering CCC—seems to be vastly increasing, despite the fact that the very early universe and the very remote future appear to be uncomfortably similar to one another. Of course they are not really similar in the sense of being nearly identical, but they are alarmingly 'similar' according to the usage of that word commonly applied in Euclidean geometry, namely that the distinction between the two seems to be basically just an enormous scale change. Moreover, any overall change of scale is essentially *irrelevant* to measures of *entropy*—where that quantity is defined by Boltzmann's marvellous formula (given in §1.3)—because of the important fact, noted at the end of §3.1, that phase-space volumes are unaltered by conformal scale changes.[3.40] Yet, the entropy *does* seem to increase, *vastly*, in our universe, through the effects of gravitational clumping. Our conundrum is to understand how these apparent facts are to square with one another. Some physicists have argued that the ultimate maximum entropy achieved by our universe will arise not from clumping to black holes, but from the Bekenstein–Hawking entropy of the *cosmological* event horizon. This possibility will be addressed in §3.5, where I argue that it does not invalidate the discussion of this section.

Let us examine more carefully the likely state of the early universe, where some appropriate condition has been imposed to kill off gravitational degrees of freedom at the Big Bang, so that gravitational entropy is low in what we find in the early universe. Do we need to take into account cosmic inflation? The reader will have realized that I am sceptical of the reality of this presumed process (§2.6), but no matter; in this discussion it makes little difference. We can either ignore the possibility of inflation, or perhaps take the view (see §3.6) that CCC provides merely a different interpretation of inflation where the inflationary phase *is* the exponentially expanding phase of the previous aeon, or else we can simply consider the situation just *following* the cosmic 'moment'—at around 10^{-32} s—when inflation is considered just to have turned off.

As I have argued at the beginning of §3.1, it is reasonable to suppose that this early-universe state (say at around 10^{-32} s) would be dominated by conformally invariant physics, and inhabited by effectively massless ingredients. Whether or not Tod's proposal, of §2.6, is correct in all detail, it seems that we do not go too far wrong in taking the early-universe state, in which gravitational degrees of freedom are indeed hugely suppressed, to be one in which a conformal stretching out would provide us with a smooth non-singular state still essentially inhabited by massless entities, perhaps largely photons. We would need to consider also the additional degrees of freedom in the *dark matter*, also taken to be effectively massless in those early moments.

At the other end of the time scale, we have an ultimate exponentially expanding de Sitter-like universe (§2.5), again largely inhabited by massless ingredients (photons). There could well be other stray material consisting, say, of stable massive particles, but the entropy would lie almost entirely within the photons. It would appear that we still do not go far wrong (appealing to the results of Friedrich cited in §3.1) if we assume that we can conformally squash down the remote future to obtain a smooth universe state not at all unlike that we obtained by conformally stretching out the situation close to the Big Bang (say at 10^{-32} s). If anything, there might well be *more* degrees of freedom activated in the stretched-out Big Bang, because in addition to the degrees of freedom perhaps activated in dark matter, Tod's proposal still allows for the presence of gravitational degrees of freedom in a non-zero (but finite) Weyl tensor **C**, rather than the requirement **C**=0 demanded by CCC (see §2.6, §3.2). But if such degrees of freedom are indeed present, this will only make our conundrum more severe, the problem to be faced being that the entropy of the very early universe is seen to be hardly smaller (if not actually larger) than that to be found in the very remote future, despite the fact that there must surely be absolutely enormous increases in entropy taking place between 10^{-32} s and the very remote future.

In order to address this conundrum properly, we need to understand the nature and magnitude of the major contributions to what we expect to be an enormous increase of entropy. At the present time, it appears that easily the major contribution to the entropy of the universe comes from huge black holes at the centres of most (or all?) galaxies. It is hard to find an accurate estimate of the sizes of black holes in galaxies generally. By their very nature, black holes are hard to see! But our own galaxy may well be fairly typical, and it appears to contain a black hole of some $4\times10^6 M_\odot$ (see §2.4), which by the Bekenstein–Hawking entropy formula provides us with an entropy per baryon for our galaxy of some 10^{21} (where 'baryon', here means, in effect, a proton or neutron, where for ease of description, I am taking baryon number to be conserved—no violation of this conservation principle having been yet observed). So let us take this figure as a plausible estimate for the current entropy per baryon in the universe generally.[3.41] If we bear in mind that the next largest contribution to the entropy appears to be the CMB, where the entropy per baryon is not more than around 10^9, we see how stupendously the entropy appears to have increased already, since the time of decoupling—let alone since 10^{-32} s—and it is the entropy of black holes that is basically responsible for this vast entropy increase. To make this more dramatic, let me write this out in more everyday notation. The entropy per baryon in the CMB is around 1 000 000 000, whereas (according to the above estimate), the current entropy per baryon is about

$$1\,000\,000\,000\,000\,000\,000\,000,$$

this being mainly in black holes. Moreover, we must expect these black holes, and consequently the entropy in the universe, to grow very considerably in the future, so that even *this* number will be utterly swamped in the far future. Thus, our conundrum takes the form of the question: how can this be squared with what has been said in the early parts of this section? What will ultimately have happened to all this black-hole entropy?

We must try to understand how the entropy will ultimately *appear* to have shrunk by such an enormous factor. In order to see where all that entropy has gone, let us recall what, indeed, is supposed to be the fate, in the very remote future, of all those black holes responsible for the vast entropy increase. According to what has been said in §2.5, after around 10^{100} years or so the holes will all have gone, having evaporated away through the process of Hawking radiation, each presumed to disappear finally with a 'pop'.

We must bear in mind that the raising of entropy due to the swallowing of material by a black hole, and also the hole's eventual reduction in size (and mass) due to its Hawking evaporation, would be fully consistent with the Second Law; not only that, but also these phenomena are direct *implications* of the Second Law. To appreciate this, in a general way, we do not need to understand the subtleties of Hawking's 1974 initial argument for the temperature and entropy of a black hole (taken to have formed in the distant past from some gravitational collapse). If we are not concerned with the *exact* coefficient $8kG\pi^2/ch$ that appears in the Bekenstein–Hawking entropy formula of §2.6, and we would be satisfied with some approximation to it, then we can find justification for the general form of black-hole entropy purely from Bekenstein's original[3.42] 1972 demonstration, which was an entirely physical argument based on the Second Law and on quantum-mechanical and general-relativistic principles, as applied to imagined experiments concerned with the lowering of objects into black holes. Hawking's black-hole surface temperature T_{BH}, which for a non-rotating hole of mass M is

$$T_{BH} = \frac{K}{M}$$

(the constant K in fact being given by $K=1/(4\pi)$), then follows from standard thermodynamic principles[3.43] once the entropy formula is accepted. This is the temperature as seen from infinity, and the rate at which a black hole will radiate is then determined by assuming that this temperature is spread uniformly over a sphere whose radius is the Schwarzschild radius (see §2.4) of the black hole.

I am stressing these points here, just to emphasize that black-hole entropy and temperature, and the process of Hawking evaporation of these strange entities, albeit of an unfamiliar character, are nevertheless very much a part of the physics of our universe, fitting in with fundamental principles that we have become accustomed to—most particularly with the *Second Law*. The enormous entropy that black holes possess is to be expected from their irreversible character and the remarkable fact that the structure of a stationary black hole needs only a very few parameters to characterize its state.[3.44]Since there must be a vast volume of phase space corresponding to any particular set of values of these parameters, Boltzmann's formula (§1.3) suggests a very large entropy. From the consistency of physics as a whole, we have every reason to expect that our present general picture of the role and the behaviour of black holes will indeed hold true—except that the eventual 'pop' at the end of the black-hole's existence is somewhat conjectural. Nevertheless, it is hard to see what else could happen to it at that stage.

But do we *really* need to believe in the pop? As long as the space-time described by the black hole remains a *classical* (i.e. non-quantum) geometry, the radiation should continue to extract mass/energy from the hole at such a rate that would cause it to disappear in a finite time—this time being $\sim 2\times 10^{67}(M/M_{\odot})^3$ years, for a hole of mass M if nothing more falls into the hole.[3.45] But how long can we expect the notions of classical space-time geometry to provide a reliable picture? The general expectation (just from dimensional considerations) is that only when the hole approaches the absurdly tiny Planck dimension l_P of $\sim 10^{-35}$m (around 10^{-20} of the classical radius of a proton) do we expect to have to involve some form of quantum gravity, but whatever happens at that very late stage, the only mass left would presumably be somewhere around that of a Planck mass m_P, with an energy content of only around a Planck energy E_P, and it is hard to see that it then could last very much longer than around a Planck time t_P (see end of §3.2). Some physicists have contemplated the possibility that the end-point might be a *stable* remnant of mass $\sim m_P$, but this causes some difficulties with quantum field theory.[3.46] Moreover, whatever the final fate of a black hole might be, its final state of existence seems to be independent of the hole's original size, and has to do with just an utterly minute fraction of the black-hole's mass/energy. There appears to be no complete agreement among physicists about the final state of this tiny remnant of a black hole,[3.47] but CCC would require that *nothing* with rest-mass should persist to eternity, so the 'pop' picture (together with the ultimate decaying away of the rest-mass of any massive particle produced in the pop) is very acceptable from CCC's point of view, and it is also consistent with the Second Law.

Yet, despite all this consistency, there is something distinctly odd about a black hole, in that the future evolution of the space-time, seemingly unique among future-evolving physical phenomena, results in an inevitable internal *space-time singularity*. Although this singularity is a consequence of *classical* general relativity (§2.4, §2.6), it is hard to believe that this classical description would have to be seriously modified by quantum-gravity considerations, until enormous space-time curvatures are encountered, where the radii of curvature of space-time begin to get down to the extremely tiny scale of the Planck length l_P (see end of §3.2). Particularly for a huge galactic-centre black hole, the place where such tiny curvature radii begin to show themselves will be an utterly minute region hugging the singularity in the classical space-time picture. The location referred to as a 'singularity' in classical space-time descriptions should really be thought of as where 'quantum gravity takes over'. But in practice this makes little difference, since there is no generally accepted mathematical structure to replace Einstein's picture of continuous space-time, so we do not ask what happens further, but merely adjoin a singular boundary of wildly diverging curvature, possibly acting in accordance with BKL-type chaotic behaviour (§2.4, §2.6).

To get a better understanding of the role of this singularity in the classical picture, we do well to examine the conformal diagram Fig. 3.13, whose two parts are basically re-drawings of Fig. 2.38(a) and Fig. 2.41, respectively. These pictures, when interpreted as strict conformal diagrams, incorporate exact spherical symmetry, which is unlikely to remain at all accurate whenever irregularities are present in the collapse. However, if we allow ourselves to assume *strong cosmic censorship* (see end of §2.5, and §2.6) right up until just before the pop,[3.48] then the singularity should be essentially *spacelike*, and the pictures of Fig. 3.13 remain qualitatively appropriate as *schematic* conformal diagrams, despite the extreme irregularity in the space-time geometry near the classical singularity.

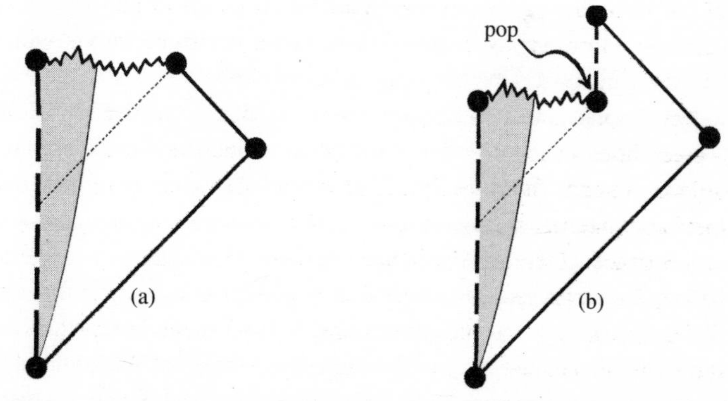

Fig. 3.13 Conformal diagrams drawn irregularly (to suggest a lack of symmetry) to indicate (a) gravitational collapse to a black hole; (b) collapse followed by Hawking evaporation. The singularity remains spacelike according to strong cosmic censorship.

The regions where one should expect quantum-gravity effects to invalidate the classical space-time picture would be very close indeed to the singularity, where space-time curvatures begin to reach the extremes where classical space-time physics can no longer be trusted. At this stage, there seems to be very little hope of adopting a standpoint like that involved in the 'crossover 3-surfaces' of CCC, in which the space-time could be extended smoothly through the singularity in order to arrive at some kind of continuation through to 'the other side'. Indeed, Tod's proposal is intended to *distinguish* the very tame singularity encountered at the Big Bang from the kind of thing—perhaps of chaotic BKL nature—that one must expect at a black hole's singularity. Despite Smolin's stimulating proposal described in §3.3 (Fig. 3.12), I see little hope of quantum gravity coming to our rescue, so as to allow us to obtain a 'bounce' of some kind, for which the emerging space-time mirrors what came in, in any direct sense, according to some kind of basically time-symmetric fundamental physical processes. If it could, then what emerges would be something of the character of the white hole of Fig. 2.46 or the mess of bifurcating white holes that we contemplated in §2.6 (contrast Fig. 3.2). Such behaviour would certainly be most unlike the kind of situation that we find in the universe with which we are familiar, and would possess nothing resembling the Second Law of our experiences.

Be that as it may, what seems to be happening—at least according to any kind of physical evolution that we seem to be able to contemplate—is that physics comes to an *end* at such regions. Or, if it does not, then it continues into some kind of universe-structure of a completely foreign character to that of which we have knowledge. Either way, the material encountering the singular region is lost to the universe we know, and it seems that any information carried by that material is also lost. But is it lost? Or can it somehow slither its way out sideways, in the diagram of Fig. 3.13(b), where quantum-gravity distortions of normal ideas of space-time geometry are somehow permitting a kind of seemingly space-like propagation that would be illegal according to the normal causality rules of §2.3? Even if so, it is hard to see that any of this information could emerge, by such means, very much before the moment of the pop, so that the vast amount of information that was contained in the material that went to form a large black hole, say of many millions of solar masses, could somehow all come flooding out just at around that one moment, and from that tiny region, that constitutes the pop. Personally, I find this very hard to believe. It seems to me to be much more plausible that the information contained in all processes whose future evolution is directed into such a space-time singularity is accordingly destroyed.

However, there is an alternative suggestion,[3.49] frequently argued for, that somehow the information has been 'leaking out' for a long time previously, encoded in what are referred to as 'quantum entanglements', that would be expressed in subtle correlations in the Hawking radiation coming from the hole. On this view, the Hawking radiation would not be exactly 'thermal' (or 'random'), but the full information that would seem to have been irretrievably lost in the singularity is somehow taken fully into account (repeated?) outside the hole. Again I have my severe doubts about any such suggestions. It would seem that, according to proposals of this kind, whatever information finds its way to the vicinity of the singularity, must somehow be 'repeated' or 'copied' as this external entanglement information, which would in itself violate basic quantum principles.[3.50]

Moreover, in his original 1974 argument,[3.51] demonstrating the presence of thermal radiation emanating from a black hole, Hawking explicitly made use of the fact that information coming in, in the form of a test wave, would have to be shared between what escapes from the hole and what falls into it. It is the assumption that the part that falls into the hole is irretrievably lost that leads to the conclusion that what comes out must have a thermal character, with a temperature that is precisely equal to what we now call the Hawking temperature. This argument depends upon use of the conformal diagram of Fig. 2.38(a), which to me makes it manifestly clear that the incoming information is indeed shared between that falling into the hole and that escaping to infinity, where that falling into the hole is lost—this being an essential part of the discussion. Indeed, for many years, Hawking himself has been one of the strongest proponents of the viewpoint that information is indeed lost in black holes. Yet, at the 17th International Conference on General Relativity and Gravitation, held in Dublin in 2004, Hawking announced that he had changed his mind and, publicly forfeiting a bet that he (and Kip Thorne) had made with John Preskill, argued that he had been mistaken and that he now believed[3.52] that the information must in fact all be retrieved externally to the hole. It is certainly my personal opinion that Hawking should have stuck to his guns, and that his earlier viewpoint was far closer to the truth!

However, Hawking's revised opinion is much more in line with what might be regarded as the 'conventional' viewpoint among quantum field theorists. Indeed, the actual destruction of physical information is not something that appeals to most physicists, the idea that information can be destroyed in a black hole in this way being frequently referred to as the 'black-hole information *paradox*'. The main reason that physicists have trouble with this information loss is that they maintain a faith that a proper quantum-gravity description of the fate of a black hole ought to be consistent with one of the fundamental principles of quantum theory known as *unitary evolution*, which is basically a time-symmetric[3.53] deterministic evolution of a quantum system, as governed by the fundamental *Schrödinger equation*. By its very nature, information cannot be lost in the process of unitary evolution, because of its reversibility. Hence the information loss that seems to be a necessary ingredient of Hawking evaporation of black holes *is*, in fact, *in*consistent with unitary evolution.

I cannot go into the details of quantum theory here,[3.54] but a brief mention of the rudimentary ideas will be important for our further discussion. The basic mathematical account of a quantum system at a particular time is provided by the *quantum state* or *wavefunction* of the system, for which the Greek letter ψ is frequently used. As mentioned above, if left to itself the quantum state ψ evolves with time according to the Schrödinger equation, this being unitary evolution, a deterministic, basically time-symmetric, continuous process for which I use the letter **U**. However, in order to ascertain what value some observable parameter q might have achieved at some time t, a quite different mathematical process is applied to ψ, referred to as *making an observation*, or *measurement*. This is described in terms of a certain operation $\mathcal{O}$ which is applied to ψ, providing us with a set of possible alternatives ψ_1, ψ_2, ψ_3, ψ_4, ..., one for each of the possible outcomes q_1, q_2, q_3, q_4, ... of the chosen parameter q, and with respective probabilities P_1, P_2, P_3, P_4, ... for these outcomes. This entire set of alternatives, with corresponding probabilities, is determined by $\mathcal{O}$ and ψ by a specific mathematical procedure. To mirror what actually appears *to happen* in the physical world, upon measurement, we find that ψ simply *jumps* to *one* of the given set of alternatives ψ_1, ψ_2, ψ_3, ψ_4, ..., say to ψ_j, where this choice appears to be completely random, but with a probability given by the corresponding P_j. This replacement of ψ by the particular choice ψ_j that Nature comes up with is referred to as the *reduction of the quantum state* or the *collapse of the wavefunction*, for which I use the letter **R**. Following this measurement, which has caused ψ to jump (to ψ_j), the new wavefunction ψ_j again proceeds according to **U** until a new measurement is made, and so on.

What is particularly strange about quantum mechanics is this very curious hybrid, whereby the quantum state's behaviour seems to alternate between these two quite different mathematical procedures, the continuous and deterministic **U** and the discontinuous and probabilistic **R**. Not surprisingly, physicists are not happy with this state of affairs, and will adopt one or another of a number of different philosophical standpoints. Schrödinger himself is reported (by Heisenberg) to have said, 'If all this damned quantum jumping were really here to stay then I should be sorry I ever got involved in quantum theory.'[3.55] Other physicists, fully appreciative of the great contribution that Schrödinger made with the discovery of his evolution equation, while agreeing with his distaste for 'quantum jumping' would nevertheless take issue with Schrödinger's standpoint that the full story of quantum evolution has not yet fully emerged. It indeed is a common view that the full story is somehow contained within **U**, together with some appropriate 'interpretation' of the meaning of ψ—and somehow **R** will emerge from all this, perhaps because the true 'state' involves not just the quantum system under consideration but its complicated environment also, including the measuring device, or perhaps because *we*, the ultimate observers, are ourselves part of a unitarily evolving state.

I do not wish to enter into all the alternatives or contentions that still thoroughly cloud the **U/R** issue, but simply state my own position, which is to side basically with Schrödinger himself, and with Einstein, and perhaps more surprisingly with Dirac,[3.56] to whom we owe the general formulation of present-day quantum mechanics,[3.57] and to take the view that present-day quantum mechanics is a *provisional* theory. This is despite all the theory's marvellously confirmed predictions and the great breadth of observed phenomena that it explains, there being no confirmed observations which tell against it. More specifically, it is my contention that the **R** phenomenon represents a *deviation* from the strict adherence of Nature to unitarity, and that this arises when gravity begins to become seriously (even if subtly) involved.[3.58] Indeed, I have long been of the opinion that information loss in black holes, and its consequent violation of **U**, represents a powerful part of the case that a strict adherence to **U** *cannot* be part of the true (still undiscovered) quantum theory of gravity.

I believe that it is this that holds the *key* to the resolution of the conundrum that confronted us at the beginning of this section. I am thus asking the reader to accept information loss in black holes—and the consequent violation of unitarity—as not only plausible, but a necessary *reality*, in the situations under consideration. We must re-examine Boltzmann's definition of entropy in the context of black-hole evaporation. What does 'information loss' at the singularity actually mean? A better way of describing this is as a *loss of degrees of freedom*, so that some of the parameters describing the phase space have disappeared, and the phase space has actually become *smaller* than it was before. This is a completely new phenomenon when dynamical behaviour is being considered. According to the normal idea of dynamical evolution, as described in §1.3, the phase space $\mathcal{P}$ is a fixed thing, and dynamical evolution is described by a point moving in this fixed space, but when the dynamical evolution involves a loss of degrees of freedom at some stage, as appears to be the case here, the phase space actually *shrinks* as part of the description of this evolution! In Fig. 3.14, I have tried to illustrate how this process would be described, using a low-dimensional analogue.

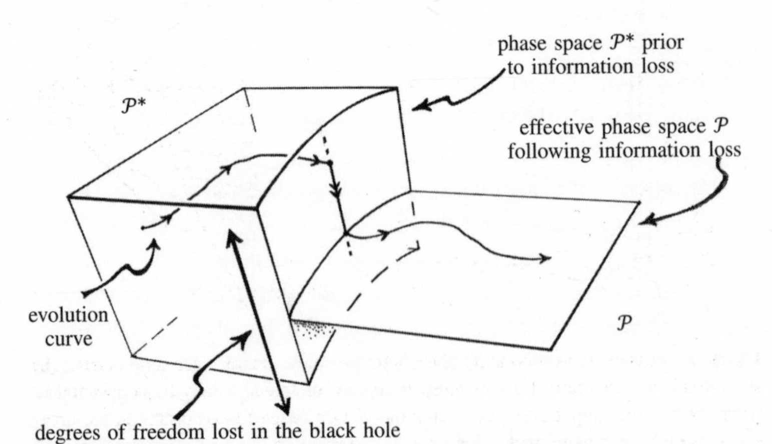

Fig. 3.14 Evolution in phase space following black-hole information loss.

In the case of black-hole evaporation, this is a very subtle process, and we should not think of this shrinking as taking place 'suddenly' at any particular time (e.g. at the 'pop'), but surreptitiously. This is all tied up with the fact that, in general relativity, there is no unique 'universal time', and this is of particular relevance in the case of a black hole, where the space-time geometry deviates greatly from spatial homogeneity. This is well illustrated in the Oppenheimer–Snyder collapse picture (§2.4, see Fig. 2.24), with final Hawking evaporation (§2.5, see Fig. 2.40 and Fig. 2.41) where in Fig. 3.15(a) and its strict conformal diagram Fig. 3.15(b) I have drawn, with unbroken lines, one family of spacelike 3-surfaces (constant time slices) where all the information lost in the hole seems to disappear at the 'instant' of the pop, whereas, using broken lines I have drawn a different family of spacelike 3-surfaces for which the information appears to go away gradually, spread out over the entire history of the black hole's existence. Although the pictures strictly refer to spherical symmetry, they still apply in a schematic way so long as strong cosmic censorship is assumed (except, of course, at the pop itself).

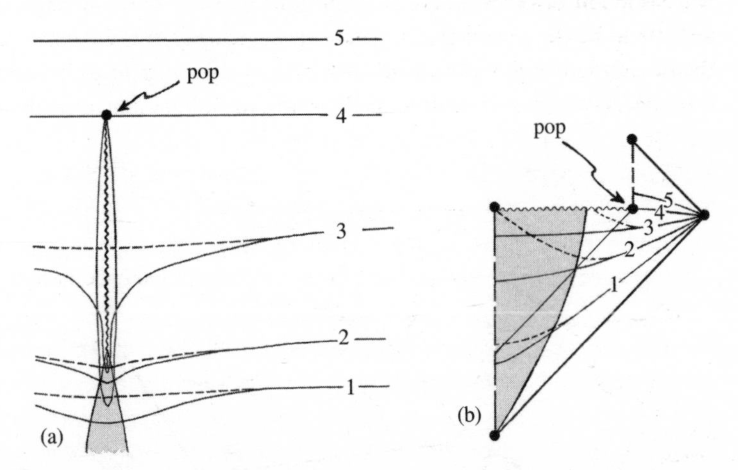

Fig. 3.15 A Hawking-evaporating black hole: (a) conventional space-time picture; (b) strict conformal diagram. Loss of internal degrees of freedom may be considered to result only as the 'pop' occurs, this being the picture suggested according to the time-slices given by unbroken lines. Alternatively, according to the time-slices given by the broken lines, the loss occurs gradually over the whole history of the black hole.

This indifference to when the information loss actually takes place serves to emphasize the fact that its disappearance has no effect on the external (thermo)dynamics, and we may well take the view that the Second Law is proceeding according to its normal practice, where the entropy continues to increase—but we must be careful about what 'entropy' notion we are referring to here. This entropy refers to *all* the degrees of freedom, including that of all the material that has fallen into the holes. The degrees of freedom referring to what has fallen in, however, will sooner or later have to confront the singularity and will, according to the considerations above, be lost to the system. By the time the black hole has disappeared with a pop, we must have radically reduced the scale of our phase space so that—as in a country experiencing currency devaluation—the phase-space volumes will count, overall, for far less than they did before, though this huge devaluation will not be noticed by the local physics continuing away from the hole in question. Because of the logarithm in Boltzmann's formula, this scaling down of the volumes would simply count as though a large constant had been subtracted from the overall entropy of the universe external to the black hole in question.

We may compare this with the discussion at the end of §1.3 where it was noted that the logarithm in Boltzmann's formula is what gives rise to the additivity of entropy for independent systems. In the foregoing discussion, the degrees of freedom swallowed and finally destroyed by the black holes play the role of the *external* part of the system under consideration in §1.3, with parameters defining the external phase space $\mathcal{X}$ which referred to the Milky Way galaxy external to the laboratory—whereas here it refers to the black holes. See Fig. 3.16. What we are now taking to be the world outside black holes, where we might envisage some experiment being performed, corresponds, in the discussion in §1.3 (Fig. 1.9), to the *internal* part of the system, defining the phase space $\mathcal{P}$. Just as the removal of degrees of freedom in the Milky Way galaxy of §1.3 (such as some of them being absorbed into the galaxy's central black hole) would make no difference whatever to entropy considerations in the experiment being performed, so also would the information destruction in black holes throughout the universe, finalized as each individually disappears with its pop, yield *no* effective violation of the Second Law, consistently with what has been emphasized earlier in this section!

Nevertheless, the phase-space volume of the universe *as a whole* would be very drastically reduced by this information loss,[3.59] and this is basically what we need for the resolution of the conundrum posed at the beginning of this section. It is a subtle matter, and there are many detailed issues of consistency to be satisfied for the reduction in phase-space volume to be adequate for what is required for CCC. In general

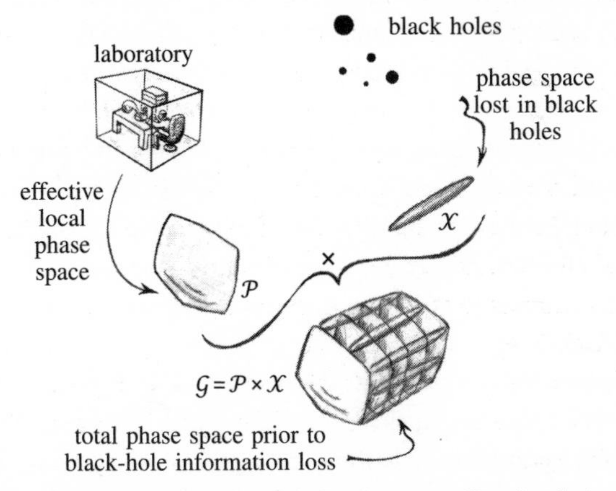

Fig. 3.16 The information loss in black holes does not affect local phase space (compare Fig. 1.9), though it contributes to the total, prior to loss.

terms, this consistency seems not unreasonable, since the overall entropy increase that our present aeon will indulge in throughout its entire history is expected to be through the formation (and evaporation) of black holes. Although it is not totally obvious to me how one calculates, with any degree of precision, the effective entropy reduction due to information loss, one could, as a good guess, estimate the Bekenstein–Hawking entropy of the maximum sizes that the black holes would have reached, had it not been for the loss in Hawking radiation, and take the total of this entropy to give the needed scale of reduction in available phase space for the start of the next aeon. Clearly there are many matters in need of more detailed study in order for us to be certain whether CCC is viable in this respect. But I can see no reason to expect that CCC will be contradicted by such considerations.

3.5 CCC and quantum gravity

The scheme of CCC provides us with a different outlook on various intriguing issues—in addition to the Second Law—that have confronted cosmology for many years. In particular, there is the question of how we are to view the singularities that arise in the classical theory of general relativity, and of how quantum mechanics enters into this picture. We find that CCC has something particular to say, not only about the nature of the Big-Bang singularity, but also about what happens when we try to propagate our physics, as we know it, into the future as far as it will go, where it apparently either terminates irretrievably at a singularity in a black hole, or else continues into the indefinite future, to be reborn, according to CCC, in the big bang of a new aeon.

Let me begin this section by examining again the situation in the very remote future, in order to raise an issue that I had left aside in the previous section. When, in §3.4, I addressed the matter of the increasing of entropy into the very remote future, I argued that, in accordance with CCC, by far the major entropy-raising processes are the formation (and congealing) of large black holes, followed by their eventual evaporation away through Hawking radiation after the CMB cools to lower than the Hawking temperature of the holes. Yet, as we have seen, CCC's requirement that the initial phase-space coarse-graining region (§1.3, §3.4) must be capable of actually matching the final one, despite the enormous increase in entropy, can be satisfied if we accept a huge 'information loss' in black holes (as Hawking originally argued for but later retracted), this allowing the phase space to become enormously 'thinned down' owing to a vast loss of phase-space dimensionality from the swallowing, and subsequent destruction, of degrees of freedom by the holes. Once the black holes have all evaporated away, we find that the zero of the entropy measure must be reset, because of this great loss of degrees of freedom, which means, in effect, that a very large number gets subtracted from the entropy value, and the allowable states in the ensuing big bang for the following aeon find themselves greatly restricted, so as to satisfy a 'Weyl curvature hypothesis', this providing the potential for gravitational clumping in the succeeding aeon.

There is, however, another ingredient to this discussion, at least in the opinion of a good many cosmologists, which I have ignored, despite its having some definite relevance to our central topic (see the end of the first paragraph of §3.4). This is the issue of 'cosmological entropy', arising from the existence of *cosmological event horizons* when $\Lambda>0$. In Fig. 2.42(a),(b), I illustrated the idea of a cosmological event horizon, which occurs when there is the *spacelike* future conformal boundary $\mathscr{I}^+$ that arises if there is a positive cosmological constant Λ. We recall that a cosmological event horizon is the past light cone of the ultimate endpoint o^+ (on $\mathscr{I}^+$) of the 'immortal' observer O of §2.5; see Fig. 3.17. If we take the view that such event horizons should be treated in the same way as black-hole event horizons, then the same Bekenstein–Hawking formula for black-hole entropy ($S_{\rm BH}=\frac{1}{4}A$; see §2.6) should be applied also to a cosmological event horizon. This gives us an ultimate 'entropy' value

$$S_\Lambda=\tfrac{1}{4}A_\Lambda,$$

in Planck units, where A_Λ is the area of spatial cross-section of the horizon in the remote future limit. In fact, we find (see Appendix B5) that this area is *exactly*

$$A_\Lambda=\frac{12\pi}{\Lambda}$$

in Planck units, so this proposed entropy value would be

$$S_\Lambda = \frac{3\pi}{\Lambda},$$

which depends solely on the value of Λ, and has nothing to do with any of the details of what has actually happened in the universe (where I assume that Λ is indeed a cosmological *constant*). In conjunction with this, if we accept the validity of the analogy, we would expect a temperature,[3.60] which is argued to be

$$T_\Lambda = \frac{1}{2\pi}\sqrt{\frac{\Lambda}{3}}$$

With the observed value of Λ, the temperature T_Λ would have the absurdly tiny value $\sim 10^{-30}$ K, and the entropy S_Λ would have the huge value $\sim 3 \times 10^{122}$.

It should be pointed out that this entropy value is far larger than what we would expect could be achieved through the formation and final evaporation of black holes in the presently observable universe as we find it, which could hardly be expected to reach more than around 10^{115}. These would be black holes in the region within our present particle horizon (§2.5). But we should ask what region of the universe is the entropy S_Λ supposed to refer to? One's first reaction might be to think that it is intended to refer to the ultimate entropy of the whole universe, since it is just a single number, precisely determined by the value of the cosmological constant Λ, and independent not only of any of the detailed activity going on within the universe, but also of the choice of eternal observer O, who provides for us the particular future end-point o^+ on $\mathscr{I}^+$. However, this viewpoint will not work, particularly because the universe might be spatially infinite, with an indefinite number of black holes within it altogether, in which case the *present* entropy of the universe could easily exceed S_Λ, in contradiction with the Second Law. A more appropriate interpretation of S_Λ might well be that it is the ultimate entropy of that portion of our universe encompassed by some cosmological event horizon—the past light cone of some arbitrarily chosen o^+ on $\mathscr{I}^+$. The material involved in this entropy would be the portion lying within the *particle* horizon of o^+ (see Fig. 3.17).

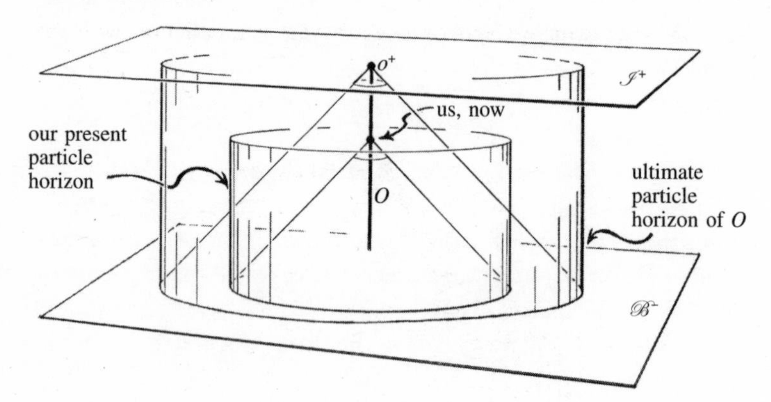

Fig. 3.17 In the current picture of our universe/aeon, our present particle horizon has a radius of roughly ⅔ of that where our ultimate particle horizon is expected to reside.

As we shall be seeing in §3.6, by the time o^+ is reached, the universe within its particle horizon should, according to the evolution of the universe that standard cosmology predicts,[3.61] then be about $(\frac{3}{2})^3 \approx 3.4$ times more than the amount of material within our *present* particle horizon, so if *that* material were all to be collected into a single black hole we would get an entropy of around $(\frac{3}{2})^6 \approx 11.4$ times 10^{124}, where 10^{124} was cited in §2.6 as a rough *upper limit* for the entropy attainable by the material within our present observable universe. Thus we get a possible black hole with an entropy of around 10^{125}. If that entropy were in principle attainable within a universe with our observed value of Λ, then we should have a gross violation of the Second Law (since $10^{125} \gg 3 \times 10^{122}$). However, if we accept the above value of T_Λ for an irreducible ambient temperature of a universe, for the observed value of Λ, then such an enormous black hole would always remain cooler than this ambient temperature, so it would never evaporate away by Hawking radiation. This still causes a problem, because we could choose o^+ to be a point on $\mathscr{I}^+$ outside this monstrous black hole, whose past light cone nevertheless encounters that hole (in the same sense that an external past light cone might ever be considered to encounter a back hole), so it seems that its entropy ought to be included —see

Fig. 3.18, and again we appear to get a vast contradiction with the Second Law.

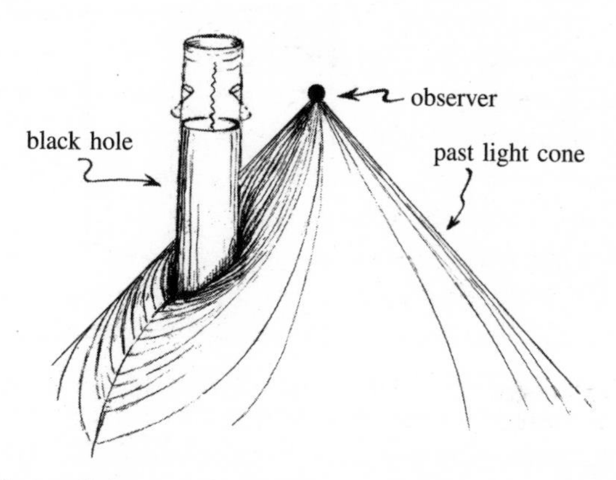

Fig. 3.18 The past light-cone of any 'observer' whether or not at $\mathscr{I}^+$, 'encounters' a black hole by *engulfing* it, rather than by intersecting its horizon.

Moreover, we have a certain leeway here, and can consider that this amount of material—around 10^{81} baryons' worth (3.4 times the 10^{80} within our presently observable universe, times about 3 because there is about this much more dark matter than baryonic matter)—can be separated into 100 separate regions of the mass of 10^{79} protons each. If each of these formed a black hole, its temperature should remain greater than T_Λ and it would evaporate away having reached an entropy of $\sim 10^{121}$. Having 100 of these, we get a total entropy of $\sim 10^{123}$ which, being larger than 3×10^{122}, seems still to violate the second Law, but not by so very much. These figures are perhaps too rough for a definitive conclusion to be deduced from them. But, in my view, they provide some initial evidence for *caution* concerning the physical interpratation of S_Λ as an actual entropy and, correspondingly, of T_Λ as an actual temperature.

I am inclined to be sceptical about S_Λ representing a *true* entropy in any case, for at least two further reasons. In the first place, if Λ really *is* a constant, so S_Λ is just a fixed number, then Λ does not give rise to any actually discernable degrees of freedom. The relevant phase space is no bigger because of the presence of Λ than it would be without it.

From the perspective of CCC, this is particularly clear, because when we match the available freedom at the $\mathscr{I}^+$ of the previous aeon with that at the $\mathscr{B}^-$ of the succeeding aeon, we find absolutely no room for the huge number of putative discernable degrees of freedom that could provide the enormous cosmological entropy S_Λ. Moreover, it seems clear to me that this comment applies also if we do *not* assume CCC, because of the remark made early in §3.4 about the invariance of the volume measure under conformal scale change.[3.62]

However, we must consider the possibility that 'Λ' is not really a constant, but is some strange kind of matter: a 'dark-energy scalar field', as favoured by some cosmologists. Then one might consider that the huge S_Λ entropy comes from the degrees of freedom in this Λ-field. Personally, I am very unhappy with this sort of proposal, as it raises many more difficult questions than it answers. If Λ is to be regarded as a varying field, on a par with other fields such as electromagnetism, then instead of calling $\Lambda\mathbf{g}$ just a separate 'Λ-term' in the Einstein field equation

$$\mathbf{E} = 8\pi\mathbf{T} + \Lambda\mathbf{g}$$

(in Planck units)—as given towards the end of §2.6—we say that there is no 'Λ-term' in the Einstein field equation, as such, but instead take the view that the Λ-field has an *energy tensor* $\mathbf{T}(\Lambda)$ which (when multiplied by 8π) is closely equal to $\Lambda\mathbf{g}$

$$8\pi\mathbf{T}(\Lambda) \cong \Lambda\mathbf{g},$$

this being now regarded as a contribution to the *total* energy tensor, which now becomes $\mathbf{T}+\mathbf{T}(\Lambda)$, and we now think of Einstein's equation as written *without* Λ-term:

$$\mathbf{E} = 8\pi\{\mathbf{T} + \mathbf{T}(\Lambda)\}.$$

But $\Lambda\mathbf{g}$ is a very strange form for $(8\pi\times)$ an energy tensor to have, being quite unlike that of any other field. For example, we think of energy as being basically the same as *mass* (Einstein's '$E=mc^2$'), and so it should have an *attractive* influence on other matter, whereas this 'Λ-field' would have a *repulsive* effect on other matter, despite its energy being positive. Even more serious, in my view, is that the *weak-energy condition* referred to in §2.4 (which is only marginally satisfied by the exact term $\Lambda\mathbf{g}$) will almost certainly be grossly violated as soon as the Λ-field is allowed to vary in a serious way.

Personally, I would say that an even more fundamental objection to referring to $S_\Lambda = \frac{3\pi}{\Lambda}$ as an actual objective entropy is that here, as opposed to the case of a black hole, there is not the physical justification of absolute *information loss* at a singularity. People have tended to make the argument that the information is 'lost' to an observer once it goes past the observer's event horizon. But this is just an observer-dependent notion; if we take a succession of spacelike surfaces like those in Fig. 3.19, we see that nothing is actually 'lost' with regard to the universe as a whole that could be associated with the cosmological entropy, since there is no space-time singularity (apart from those already present inside individual black holes).[3.63] Moreover, I am not aware of any clear *physical* argument to justify the entropy S_Λ, like the Bekenstein argument for black-hole entropy alluded to earlier in this section.[3.64]

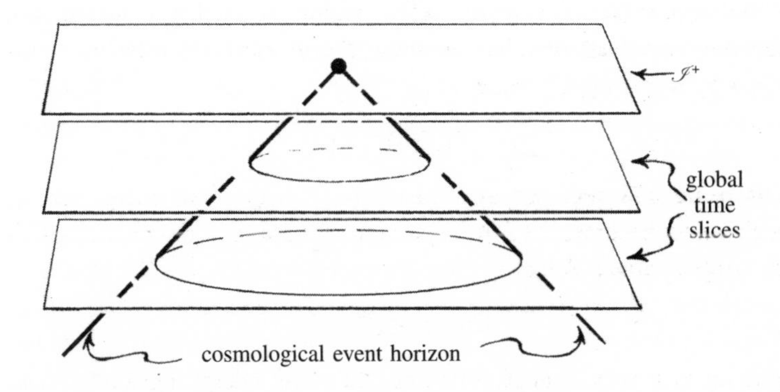

Fig. 3.19 For a cosmological event horizon there is no information loss (unlike the case of a black hole), as is evident from the all-encompassing nature of a family of global time-slices.

Perhaps my difficulty with this is made clearer in the case of the cosmological 'temperature' T_Λ since this temperature has a strongly observer-dependent aspect to it. In the case of a black hole, the Hawking temperature is provided by what is called the 'surface gravity' which has to do with the accelerating effect felt by an observer supported in a stationary configuration close to the hole, where 'stationary' refers to the relation between the observer and a reference frame held fixed at infinity. If the observer falls freely into the hole, on the other hand, then the local Hawking temperature would *not* be felt.[3.65] The Hawking temperature thus has this subjective aspect to it, and may be regarded as an instance of what is referred to as the *Unruh effect* that a rapidly accelerating observer would feel even in flat Minkowski space $\mathbb{M}$. When we come to consider the cosmological temperature of de Sitter space $\mathbb{D}$, we would expect, by the same token, that it would be an *accelerating* observer who should feel this temperature, not one who is in free fall (i.e. in geodesic motion; see end of §2.3). An observer moving freely in a de Sitter background would be *unaccelerated* in these terms, and, so it seems, should *not* experience the temperature T_Λ.

The main argument for cosmological entropy seems to be an elegant but purely formal mathematical procedure based upon analytic continuation (§3.3). The mathematics is certainly enticing, but objections can be raised to its general relevance since, technically, it applies only to exactly symmetrical space-times (like de Sitter space $\mathbb{D}$).[3.66] Again there is the subjective element of the observer's state of acceleration, arising because $\mathbb{D}$ possesses many different symmetries, corresponding to different states of observer acceleration.

This issue is brought into better focus if we look more carefully at the Unruh effect, within Minkowski space $\mathbb{M}$. In Fig. 3.20, I have tried to indicate a family of uniformly accelerating observers—referred to as *Rindler observers*[3.67]—who, according to the Unruh effect, would experience a temperature (extremely tiny, for any achievable acceleration) even though they move through a complete vacuum. This arises through considerations of quantum field theory. The future 'horizon' $\mathcal{H}_0$ for these observers associated with this temperature is also shown, and we may well take the view that there should be an entropy associated with $\mathcal{H}_0$, for consistency with this temperature and with the Bekenstein–Hawking discussion of black holes. Indeed, if we imagine what goes on in a small region in the close neighbourhood of the horizon of a very large black hole, then the situation would be very closely approximated by that shown in Fig. 3.21, where $\mathcal{H}_0$ locally coincides with the black-hole horizon and where the Rindler observers would now be the 'observers supported in a stationary configuration close to the hole' considered above. These observers are the ones who 'feel' the local Hawking temperature, whereas an observer who falls freely directly into the hole, being analogous to an inertial (unaccelerated) observer in $\mathbb{M}$, would not experience this local temperature. The entire entropy associated with $\mathcal{H}_0$ would, however, have to be *infinite*, if we carry this picture in $\mathbb{M}$ right out to infinity, which illustrates the fact that the full discussion of black-hole entropy and temperature actually involves some non-local considerations.

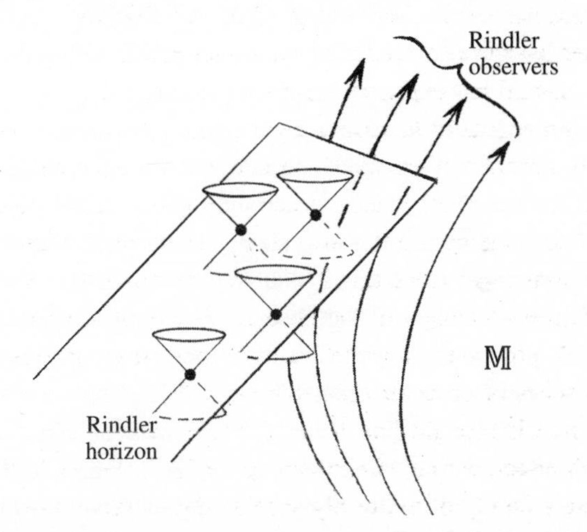

Fig. 3.20 Rindler (uniformly accelerating) observers, feeling the Unruh temperature.

A cosmological event horizon $\mathcal{H}_\Lambda$ arising when $\Lambda > 0$, as considered above, has a strong resemblance to a Rindler horizon $\mathcal{H}_0$.[3.68] Indeed, on taking the limit $\Lambda \to 0$, we find that $\mathcal{H}_\Lambda$ actually *becomes* a Rindler horizon—but now *globally*. This would be consistent with the entropy formula $S_\Lambda = 3\pi/\Lambda$ leading to $S_0 = \infty$, but it also leads us to question an assignment of objective reality to this entropy, since this *infinite* entropy seems to make little objective sense in the case of Minkowski space.[3.69]

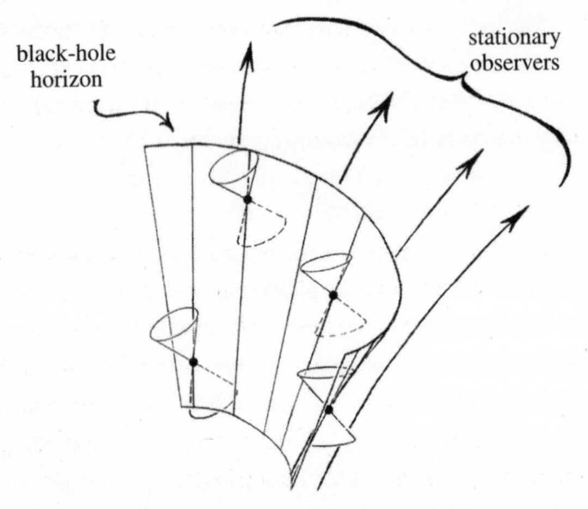

Fig. 3.21 Observers supported in a stationary configuration near a black hole's horizon feel a strong acceleration and the Hawking temperature. The situation, locally, is just like that of Fig. 3.20.

I believe that it has been worth raising these matters at some length here, because an assignment of a temperature and entropy to the vacuum is an issue of *quantum gravity* that is deeply related to the concept referred to as 'vacuum energy'. According to our current understandings of quantum field theory, the vacuum is not something totally devoid of activity, but consists of a seething bustle of processes going on at a very tiny scale, where what are called *virtual* particles and their anti-particles momentarily appear and disappear in 'vacuum fluctuations'. Such vacuum fluctuations, at the Planck scale l_P, would be expected to be dominated by *gravitational* processes, and the performing of the necessary calculations that would be needed for obtaining this vacuum energy is something far beyond the scope of currently understood mathematical procedures. Nevertheless, general arguments of symmetry, to do with the requirements of relativity, tell us that a good overall description of this vacuum energy ought to be by an energy tensor $\mathbf{T}_v$ of the form

$$\mathbf{T}_v = \lambda \mathbf{g}$$

for some λ. This looks like exactly the kind of energy term $\mathbf{T}(\Lambda)$ that would be provided by a cosmological constant, as we have seen above, so it is frequently argued that a natural interpretation of the cosmological constant is that it *is* this vacuum energy, where

$$\lambda = (8\pi)^{-1}\,\Lambda.$$

The point of view would appear to be to regard the 'degrees of freedom' responsible for the large cosmological entropy S_Λ to be those in the 'vacuum fluctuations'. These are not what I have referred to above as 'discernable' degrees of freedom since, if they count at all towards phase-space volume, they do this *uniformly* throughout space-time, and constitute merely a *background*, to which normal physical activity going on within the space-time appears not to contribute.

Perhaps even more seriously, a trouble with this interpretation appears to be that when attempted calculations are made for obtaining the actual value of λ, the answer comes out as

$$\lambda = \infty, \text{ or } \lambda = 0, \text{ or } \lambda \approx t_P^{-2},$$

t_P being the Planck time, see §3.2. The first of these answers is the most honest (and a common kind of conclusion that the *direct* application of the rules of quantum field theory tends to yield!), but it is also the most wrong. The second and the third are basically guesses as to what the answer *should* come out as after the application of one or another of the standard procedures of 'getting rid of infinities' is applied (such procedures, when applied with appropriate skill, often providing superbly accurate answers in *non*-quantum-gravity circumstances). The answer $\lambda = 0$ seems to have been the favoured one so long as it had been believed that $\Lambda = 0$ fitted the observational facts. But since the supernova observations referred to in §2.1 indicated that it is more probable that $\Lambda > 0$, and later observations supported this conclusion, a non-zero value for λ has become favoured. If the cosmological constant really *is* vacuum energy in this sense of *gravitational* 'quantum fluctuations', then the only scale available is the Planck scale, which is why it seems that t_P (or equivalently l_P), or some reasonably small multiple of it, *ought* to provide the needed scale for λ. For dimensional reasons, λ has to be the inverse square of a distance, so a rough answer $\lambda \approx t_P^{-2}$ is to be expected. However, as we have seen in §2.1, the observed value of Λ is more like

$$\Lambda \approx 10^{-120}\, t_P^{-2}$$

so something is clearly seriously wrong, either with this interpretation ($\lambda = \Lambda/8\pi$) or with the calculation!

Our understanding of these matters has not settled down to being free of contention, so it is perhaps of some interest to see what CCC has to say about them. The physical status of S_Λ and T_Λ does not *crucially* affect CCC, because even if the entropy S_Λ and the temperature T_Λ are to be regarded as physically 'true', this would not need to alter the picture presented by CCC. No black hole, that we expect to arise in the universe we know, would reach anything like the size at which T_Λ would seriously affect its evolution. As for S_Λ, it does not really appear to help with the conundrum of §3.4, since the issue there concerned *discernable* degrees of freedom (i.e. degrees of freedom that relate to actual dynamical processes) and simply introducing an 'entropy' with the fixed value $3\pi/\Lambda$ does not really change anything. We can simply ignore it, since it seems to play no role in the dynamics, and even if considered 'real' it appears to correspond to no physically discernable degrees of freedom. Either way, my personal position will be to ignore *both* S_Λ and T_Λ and to proceed without them.

The scheme of CCC *does*, on the other hand, provide a clear, but unconventional, perspective on how quantum gravity would affect the classical space-time singularities. The inevitability of space-time singularities in classical general relativity (§2.4, §2.6, §3.3) had led physicists to turn to *some* form of quantum gravity, in order to understand the physical consequences of the extraordinarily large space-time curvatures that are expected to arise in the vicinity of such singularities. But there has been very little agreement on how quantum gravity might alter these classically singular regions. There is, indeed, very little agreement about what 'quantum gravity' actually ought to *be*, in any case.

Nevertheless, theorists had learned to take the view that, so long as radii of space-time curvature are very large in comparison with the Planck length l_P (see §3.2), then a reasonably 'classical' picture of space-time can be maintained, perhaps only with tiny 'quantum corrections' to the standard classical equations of general relativity. But when space-time curvatures get extremely large, the radii of curvature getting down to the absurdly tiny scale of l_P (of some 20 orders of magnitude smaller than the classical radius of a proton), then even the standard picture of a smoothly continuous space would seem to have to be completely abandoned, and replaced by something radically different from the smooth space-time picture that we are used to.

Moreover, as had been argued strongly by John Wheeler and others, even the ordinary closely flat space-time of our experiences, if it were to be examined at the minute Planck scale, would be found to have a turbulent chaotic character, or perhaps a discrete granular one—or have some other kind of unfamiliar structure better described in some other way. Wheeler presented the case for quantum effects of gravity causing the space-time at the Planck level to curl up into topological complications that he viewed as a kind of 'quantum foam' of 'wormholes'.[3.70] Others have suggested that some kind of discrete structure might manifest itself (like entangled, knotted 'loops',[3.71] spin foams,[3.72] lattice-like structure,[3.73] causal sets,[3.74] polyhedral structure,[3.75] etc.[3.76]), or that a mathematical structure, modelled on quantum-mechanical ideas, referred to as 'non-commutative geometry'[3.77] might become relevant, or that higher-dimensional geometry might play a role, involving string-like or membrane-like ingredients,[3.78] or even that space-time itself might fade away completely, where our normal macroscopic picture of space-time arises only as a useful notion derived from a different more primitive geometric structure (as happens with 'Machian'[3.79] theories and with 'twistor' theory[3.80]). It is clear from this multitude of very different alternative suggestions that there is no agreement whatsoever as to what might actually be going on in 'space-time' at the Planck scale.

However, according to CCC, we find at the Big Bang something very different from such wild or revolutionary suggestions. We are provided with a much more *conservative* picture, were we have a perfectly smooth space-time, differing from that of Einstein only in that there is no conformal scaling provided, and where time evolution can be treated by conventional mathematical procedures. In CCC, the singularities occurring deep within black holes, on the other hand, would have a very different kind of structure from the Big-Bang singularity and we would have to consider some kind of exotic information-destroying physics which might indeed have to incorporate quantum-gravity ideas differing very much from the notions of space-time used in today's physics, and which might well have to incorporate some wild or revolutionary idea among those mentioned above.

For many years, it has been my own view that these two different singular ends of time appear to have very distinct characters. This is in keeping with the Second Law, where for some reason gravitational degrees of freedom would have to be greatly suppressed at the initial end, though not at the final end. I had always found it extremely puzzling why quantum gravity should appear to treat these two occurrences of space-time singularities in so different a way. Yet I had imagined, in accordance with what seems now to be the prevailing view, that it should be some form of *quantum gravity* that governs the kind of geometrical structure that we find close to *both* of these types of singular space-time geometry. However, apparently at variance with the common view, I had taken the position that the true 'quantum gravity' must be a grossly time-asymmetric scheme, involving whatever modification to the standard present-day rules of quantum mechanics might be required—in accordance with aspirations I have mooted, towards the end of §3.4.

What I had not anticipated, before turning to the point of view that CCC provides, is that the Big Bang should be treated as part of an essentially *classical* evolution, in which deterministic differential equations like those of standard general relativity govern behaviour. The question was: how could CCC escape the conclusion that enormous space-time curvatures, with radii down to the level of the Planck scale l_P near the Big Bang, ought to imply that *quantum gravity* enters the scene, with all the chaos this entails? CCC's answer is that there is curvature and there is curvature; or, to be more precise, there is Weyl curvature **C** and Einstein curvature **E** (the latter being equivalent to Ricci curvature; see §2.6 and Appendix A). The point of view of CCC is to agree that when radii of curvature approach the Planck scale, the madness of quantum gravity (whatever it is) must indeed begin to take over, but the curvature in question must be *Weyl* cuvature, as described by the conformal curvature tensor **C**. Accordingly, the radii of curvature involved in the Einstein tensor **E** can become as small as they like, and the space-time geometry will still remain essentially classical and smooth so long as the Weyl cuvature radii are *large* on the Planck scale (Fig. 3.22).

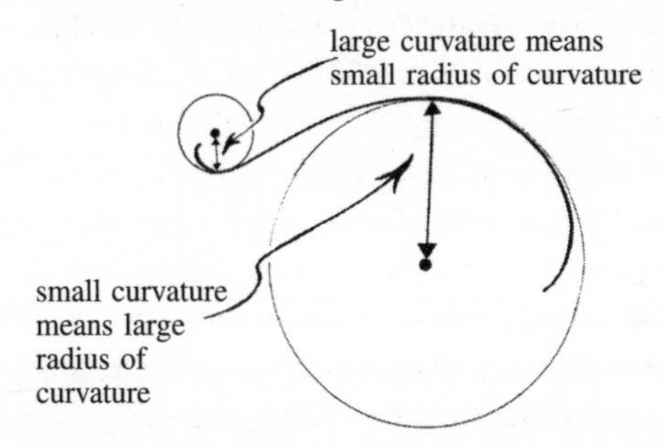

Fig. 3.22 A 'radius of curvature' is a reciprocal measure of curvature, which is small when curvature is large and large when curvature is small. Quantum gravity is commonly argued to become dominant when space-time curvature radii approach the Planck length, but CCC maintains this applies only to *Weyl* curvature.

In CCC we find that $\mathbf{C}=0$ at the Big Bang (whence *infinite* radii of Weyl curvature), so we are justified in thinking that essentially classical considerations should suffice. Thus, the detailed nature of the big bang of each aeon is completely determined by what happened in the remote future of the prior aeon, and this should lead to observational consequences, some being considered in §3.6. Here, classical equations continue the evolution of the massless fields that were present in the very remote future of the immediately preceding aeon into the big bang of the next. On the other hand, currently standard approaches to the very early universe assume that *quantum gravity* should be what determines behaviour at the Big Bang. In essence, this is the kind of way (though in terms of the 'inflaton field') that inflationary cosmology would decree how the slight deviations in the CMB temperature (of a few parts in 10^5) over the sky come about, initially, from 'quantum fluctuations'. However, CCC provides a completely different perspective on this, as we shall be seeing in the next section.

3.6 Observational implications

The question I now wish to address is whether we can find any specific evidence either for or against the actual validity of CCC. It might have been thought that any evidence concerning a putative 'aeon' existing prior to our Big Bang must be well beyond any observational access, owing to the absolutely enormous temperatures arising at the Big Bang that would seem to obliterate all information, thereby separating us from all that supposed previous activity. We should bear in mind, however, that there has to be an extreme organization present in the Big Bang, as a direct implication of the Second Law, and the arguments of this book point to this 'organization' having the character that allows our Big Bang to be extended conformally to an aeon prior to ours, this extension being governed by a very specific deterministic evolution. Accordingly, we may hope that there is a sense in which we might actually be able to 'see' through to that earlier aeon!

We must ask what particular features of the remote future of an aeon prior to ours could possibly be observable to us. One thing we can be sure about, if CCC is right, is that the overall spatial geometry of our own aeon must match that of the previous one. If the previous aeon were spatially finite, for example, then so must be our own. If that earlier aeon accorded, on a large scale, with a Euclidean spatial 3-geometry ($K=0$), then that would apply also to ours, and if it had a hyperbolic spatial geometry ($K<0$), then our own would be hyperbolic also. All this follows because the spatial geometry, overall, is determined by that of the crossover 3-surface, the geometry of this 3-surface being common to both of the aeons that it bounds. Of course, this provides us with nothing new, of observational value, because we have no independent information about the overall spatial geometry of the previous aeon.

On a somewhat smaller scale, however, matter distributions might rearrange themselves throughout the progress of each aeon, according to some perhaps complicated—but in principle comprehensible—dynamical processes. The *ultimate* behaviour of these matter distributions, taking the form of massless radiation (in accordance with CCC's §3.2 requirements), can then leave its signature on the crossover 3-surface, and then perhaps be readable in subtle irregularities in the CMB. Our task would be to try to ascertain what, in this regard, would be the most important processes taking place in the course of the previous aeon, and to try to decipher the signals hidden in such tiny irregularities in the CMB.

To be able to interpret signals of this kind, we would need to have a good understanding of the phenomena that would be likely to cause them. For this, we would need to look carefully at the dynamical processes that might be involved in the previous aeon, and also at how things might propagate from one aeon to the next. However, in order to come to any reasonably clear conclusions about the detailed nature of the previous aeon, it will be of help to us if we may assume that it was, in a general way, essentially like our own. Then we can take it that the aeon prior to ours would have behaved closely in accord with the kind of behaviour that we see in the universe around us, and with the general way that we expect it to evolve far into the future.

Most evidently, we would expect that there should have been an exponential expansion in the remote future of the previous aeon, where we are supposing that a positive cosmological constant dominated the behaviour of that aeon in its very remote future, as appears to be the case for our own (if we take Λ to be a constant). The resulting exponential expansion of that earlier aeon would bear a tantalizing similarity to the supposed inflationary phase of the currently favoured picture of the very early history of the universe, although this currently conventional picture has the exponential expansion taking place between around 10^{-36} s and 10^{-32} s in our *own* aeon (see §2.1, §2.6), closely following the Big Bang itself. On the other hand, CCC would place this 'inflationary phase' *before* the Big Bang, identifying it with the exponential expansion of the remote future of the previous aeon. In fact, as mentioned in §3.3, an idea of this nature was put forward by Gabriele Veneziano in 1998[3.81] although his scheme depended heavily on ideas from string theory.

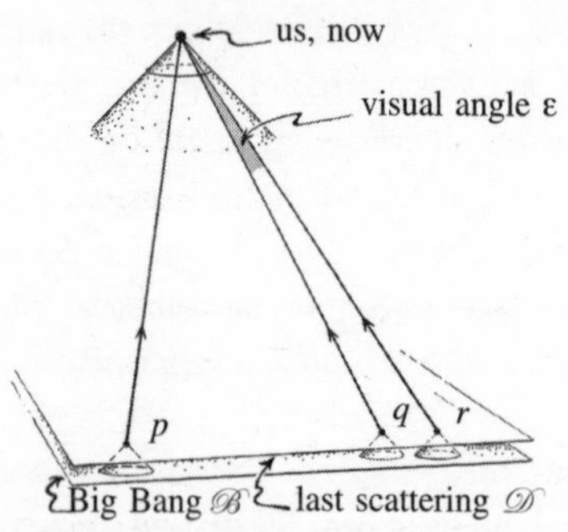

Fig. 3.23 Standard (pre-inflationary) cosmologies could imply that points in the CMB sky, farther apart than that given by $\varepsilon = 2°$ in the figure, should not be correlated (since the past light cones of q and r do not intersect), whereas such correlations are observed up to ~60°, as with the points like p and r.

One important aspect of this general idea is that two key pieces of observational evidence that have appeared to provide crucial support for the now-standard picture of inflationary cosmology, as discerned from the slight temperature variations seen in the CMB, appear also to be addressed by pre-Big-Bang theories of this nature. One of these is that there are observed *correlations* in the temperature variations in the CMB over angles in the sky (up to about 60°, in fact) that would be inconsistent with the standard cosmologies of the Friedmann or Tolman type (§2.1, §3.3), if the Big Bang itself is taken to be inherently free of correlations. This inconsistency is shown in the schematic conformal diagram of Fig. 3.23, where we see that the surface of last scattering $\mathscr{D}$, (decoupling; see §2.2) occurs much too close to the Big-Bang 3-surface $\mathscr{B}^{-}$ for effects that are seen from our vantage point to be more than about 2°

apart in the sky ever to have been in causal contact. This assumes that all such correlations arise from processes occurring *after* the Big Bang, and the different points of $\mathscr{B}^-$ are in fact completely uncorrelated. Inflation is able to achieve such correlations because the 'inflationary phase' increases the separation between $\mathscr{B}^-$ and $\mathscr{D}$ in a conformal diagram,[3.82] so that much larger angles seen from our vantage point are brought into causal contact; see Fig. 3.24.

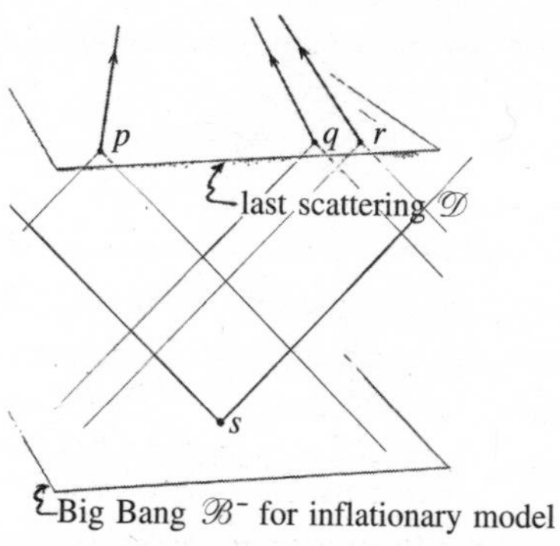

Fig. 3.24 An effect of inflation is to increase the separation between $\mathscr{D}$ and $\mathscr{B}^-$, so that the correlations of Fig. 3.23 can occur.

The other key piece of observational evidence, seeming to give powerful support for inflation, is that the initial density fluctuations—giving rise to temperature fluctuations in the CMB—appear to be *scale-invariant*, over a very broad range. The explanation from inflationary cosmology is that there were initial completely random irregularities—of the nature of initially tiny *quantum fluctuations* in the 'inflaton field' (§2.6)—very soon after the Big Bang, and that the inflationary exponential expansion then took over, expanding out these irregularities to an enormous degree, these finally being realized[3.83] in actual density irregularities in the (mainly dark) matter distribution. Now, an exponential expansion is a self-similar process, so one can imagine that, if there is randomness about how the initial fluctuations are distributed in space-time, then the result of the exponential action on these fluctuations will be a distribution with a certain *scale invariance*. In fact, long before the inflationary scheme was put forward, it had been proposed by E.R. Harrison and Y.B. Zel'dovich, in 1970, that the observed departures from uniformity in the early distribution of material in the universe could be explained if it were assumed that the initial fluctuations are indeed *scale invariant*. Not only had inflation given a rationale for this supposition, but analysis of subsequent observations of the CMB confirmed a close scale invariance over a much greater range than before, this lending some considerable support to the inflationary idea, particularly since it had been hard to see what other kind of explanation could give a theoretical basis to this observed scale invariance.

Indeed, if one is to reject the inflationary picture, then *some* alternative explanation needs to be found of both the scale invariance and the correlations beyond the horizon size in the initial density irregularities. In CCC (as in the earlier Veneziano scheme) these two points are dealt with by, in effect, displacing the inflationary phase of the universe from occurring at a moment just following the Big Bang to a phase of expansion *preceding* the Big Bang, as described above. Since we still have an effectively self-similar expanding universe phase, just as with inflation, it may be expected that this could lead to density fluctuations that have a scale-invariant nature. Moreover, correlations outside the horizon scale of the Friedmann or Tolman models are again to be expected, but now these correlations are set up through events that took place in the aeon prior to our own. See Fig. 3.25.

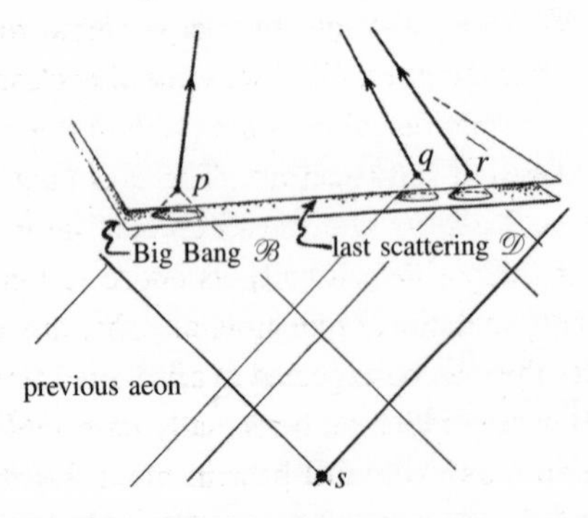

Fig. 3.25 In CCC, the required correlations of Fig. 3.23 can result from activities in the prevous aeon.

In order to be more explicit about what these events are likely to be, according to CCC, we must try to understand what are likely to be the most relevant processes taking place in the aeon prior to our own. Before we can go into much detail about this, there is a particularly big question mark that we must address. For there is the possibility, remarked upon in §3.3, that we shall have to take seriously: John A. Wheeler's suggestion that the basic constants of Nature might *not* have had precisely the same values in the previous aeon as they have in our own. The most obvious (and simplest) such possibility would be that the large number N, referred to towards the end of §3.2, which in our aeon takes a value $N \approx 10^{20}$ might, in the previous aeon, have taken some other value. There are, of course, two sides to this issue. It would certainly make life easier if we can just *assume* that fundamental numerical constants such as N had the same value in the previous aeon as in ours, or that the observations would be insensitive to (reasonable) alterations in the values of such numbers. But, on the other hand, if there are clearly distinguishable effects that changing a number such as N might have, then there is the potentially exciting possibility of actually *ascertaining* whether or not such a number might be fundamentally constant (perhaps being in principle mathematically calculable) or whether it actually *does* change from aeon to aeon, possibly in a specific mathematical way that could itself be subject to observational test.

A subsidiary set of question marks relate to our expectations about the evolution of our *own* aeon into the very remote future. Here, the requirements and expectations for CCC are somewhat clearer. Specifically, Λ must indeed be a cosmological *constant*, with our aeon continuing in its exponential expansion until eternity. The Hawking evaporation of black holes must be a *reality* and must continue until every hole has wasted away, having deposited virtually its entire rest-energy into low-energy photons and gravitational radiation, and that this will occur even for the largest holes that can be expected to arise, until finally they disappear. Might this Hawking radiation be actually *detectable* if it occurs in the aeon previous to ours? We must bear in mind that the entire mass-energy of a black hole, no matter how vast it might initially be, would ultimately have to be deposited in this low-frequency electromagnetic radiation. This energy would ultimately find its way to the crossover surface and leave its subtle imprint on the CMB of our own aeon. It is not at all out of the question, if CCC is right, that this information could eventually be teased out of the tiny irregularities in the CMB. This would be most remarkable, if so, since the Hawking radiation in our own aeon would normally be regarded as being such an absurdly tiny effect that it would be completely unobservable!

A more unconventional implication of CCC is that the rest-masses of all particles ought eventually to die away over the vast stretches of eternity (§3.2), so that in the asymptotic limit all surviving particles, including charged ones, become massless. The decaying away of rest-mass would be a universal feature of massive particles, according to this scheme, so one might imagine that it should be an observable effect. However, at the present stage of understanding, no prescription of the rate at which mass should decay away has been provided by the scheme. The decay rate might well be extremely slow so it would be hard to maintain that the fact that no such decay has yet been observed represents any evidence against this aspect of CCC. One point that is worth making here is that if all different types of particles have mass-decay rates that are closely in proportion, then the effect would appear as a very slow weakening of the gravitational constant. As of 1998,[3.84] the best experimental limit on any decay rate for the gravitational constant is that it would have to be less than about 1.6×10^{-12} per year. However, we must bear in mind that a time scale of 10^{12} years is small beer indeed, compared with the time periods of at least 10^{100} years that need to be considered to allow time for all black holes to disappear. At the time of writing, I am not aware of any clear-cut observational proposal that would seriously test the aspect of CCC that demands the ultimate decaying away of rest-mass.

There is, however, one clear implication of CCC that it ought to be possible to settle by an appropriate analysis of the CMB. The effect in question is gravitational radiation from very close encounters between extremely massive black holes (primarily those in galactic centres). What would be the result of such encounters? If the holes pass each other particularly closely, it would be expected that each would deflect the motion of the other sufficiently violently for there to be a burst of gravitational radiation that could carry away a significant amount of energy from the pair, and their relative motions would be appreciably reduced. If the encounter were extremely close, then they might well capture each other in orbits about one another, which become tighter and tighter through energy loss in gravitational waves, resulting in a huge total energy loss in this way, until they swallow each other up to form a single black hole. In extreme cases, this single hole could be the result of a direct impact, the resulting hole being initially grossly distorted before the hole settles down via gravitational radiation. In either case, there would be an enormous emission of gravitational waves that would be likely to carry away a not inconsiderable proportion of the huge combined mass of the two holes.

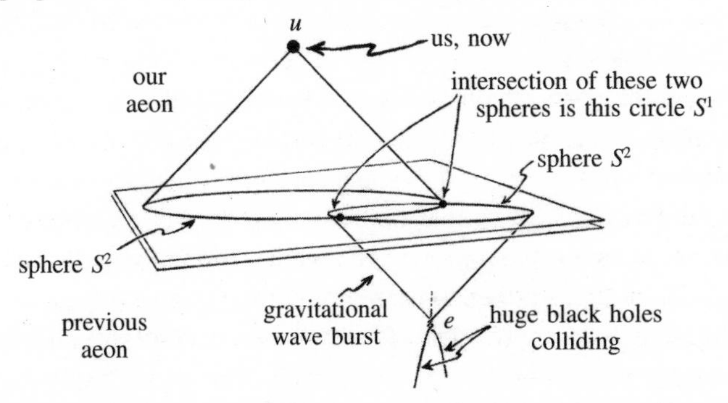

Fig. 3.26 Encounters betwen huge black holes in the prevous aeon would result in significant bursts of gravitational radiation. This should appear as a circle of enhanced or diminished temperature (depending on the overall geometry) in the CMB sky.

On the kind of time-scale that we are concerned with here, this entire burst of gravitational waves would be virtually instantaneous. In the absence of large further distorting effects throughout the universe, this radiation would be essentially contained within a thin almost spherical shell spreading out forever, from the point of encounter e, with the speed of light. In terms of a (schematic) conformal picture (Fig. 3.26) this burst of energy would be represented as an outward light cone $\mathscr{C}^+(e)$ extending from e to $\mathscr{I}^\wedge$ (where $\mathscr{I}^\wedge$ is the '$\mathscr{I}^+$' of the previous aeon to ours). Although it might be thought that this radiation would eventually become indefinitely attenuated, so as to be totally insignificant when ultimately $\mathscr{I}^\wedge$ is reached, if we look at the situation in the right way we find that this is not really the case. We recall from §3.2 that the gravitational field can be described by a $[{}^{0}_{4}]$-tensor **K**, satisfying a conformally invariant wave equation $\nabla\mathbf{K}=0$. Since this wave equation is indeed conformally invariant, we can regard **K** as propagating in the space-time depicted in Fig. 3.26, where we can regard the future boundary $\mathscr{I}^{\wedge}$ as just an ordinary spacelike 3-surface. The wave reaches $\mathscr{I}^{\wedge}$ within a finite period, and **K** has a finite value there which can be estimated from the geometry of Fig. 3.26.

Now, because of the relation between **K** and the conformal tensor **C** in the conformal metric scaling that we would use for Fig. 3.26 (the '$\hat{\mathbf{C}}=\Omega\hat{\mathbf{K}}$' of §3.2), we find that the conformal tensor **C** reaches the value *zero* at $\mathscr{I}^{\wedge}$, but it has a non-zero normal derivative across $\mathscr{I}^{\wedge}$ (see Fig. 3.27; compare with Fig. 3.6). From the arguments of Appendix B12, we find that the presence of this normal derivative has two direct effects. One of these is to influence the *conformal geometry* of the crossover surface ($\mathscr{I}^{\wedge}/\mathscr{B}^{-}$), via a conformal curvature quantity known as the 'Cotton–York' tensor, so that we cannot expect the spatial geometry of the succeeding aeon (our own) to be exactly of FLRW type at the moment of the Big Bang, but there must be slight irregularities. The second, and more immediately observable effect, would be to give the ϖ-field material—argued, in §3.2, to be the initial phase of new *dark matter*—a significant 'kick' in the direction of the radiation; see Fig. 3.27.

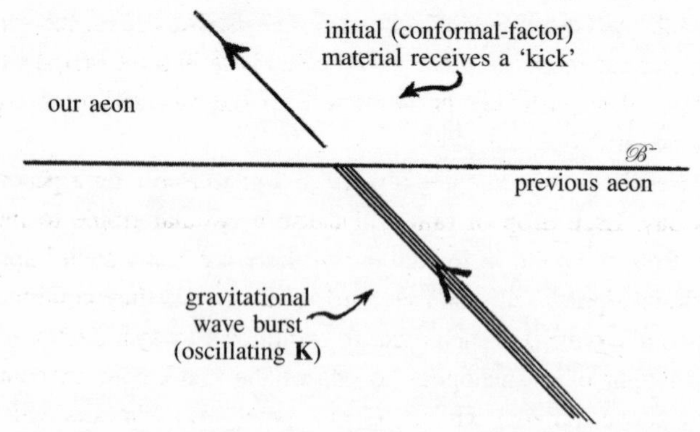

Fig. 3.27 When the gravitational wave burst encounters the crossover 3-surface, it gives the initial material of the succeeding aeon a 'kick' in the direction of the wave.

If the point u represents our present location in the space-time, then the past light cone $\mathscr{C}^-(u)$ of u represents that part of the universe that we can directly 'see'. The intersection of $\mathscr{C}^-(u)$ with the decoupling surface $\mathscr{D}$ thus represents what can be directly observed in the CMB, but since in a strict conformal representation $\mathscr{D}$ is very close (about 1% of the total height of the entire aeon, in the picture) to the crossover surface $\mathscr{B}^-$, we do not go too far wrong[3.85] if we think of this as the intersection of $\mathscr{C}^-(u)$ with $\mathscr{B}^-$. Ignoring any effects of non-uniformity of matter density within our own aeon, this will be a geometrical *sphere*. The future light cone $\mathscr{C}^+(e)$ of e will also meet $\mathscr{I}^{\wedge}$ $(=\mathscr{B}^-)$ in a geometrical sphere, assuming that we may ignore density non-uniformity in this *previous* aeon. Thus, the part of the radiation from the black-hole encounter at e that is visible directly to us through its effect on the CMB will be the intersection of these two spheres on $\mathscr{B}^-$, this intersection being a geometrically precise *circle* C, where I am here ignoring the slight difference between the 3-surfaces $\mathscr{B}^-$ and $\mathscr{D}$.

The 'kick' that the impulse of energy-momentum that the gravitational wave burst will impart on the (presumed) primordial dark matter will have a component in our direction that could be towards us or it could be away from us, depending on the geometrical relation between u, e, and the crossover surface. This effect of being towards us or away from us would be the same all around the entire circle C. Thus, we expect that for each such black-hole encounter in the previous aeon, for which these two spheres intersect, there would be a circle in the CMB sky that contributes either positively or negatively to the background average CMB temperature over the sky.

For a useful analogy, imagine a pond in a gentle rain, on a peaceful windless day. Each drop of rain will cause a circular ripple to move outwards from the point of impact, but if there are many such impacts the individual ripples will soon be hard to discern as they continually move outwards to overlap each other in complicated ways. Each impact is to be thought of as analogous to one of the black-hole encounters envisaged above. After a while, the rain peters out (the analogue of the black holes finally disappearing through Hawking evaporation), and we are left with a random-looking pattern of ripples, and from a photograph of such a pattern it would be hard to ascertain that it had been produced in this way. Nevertheless, if the appropriate statistical analysis is performed on this pattern, it ought to be possible (if the rain had not continued for too long) to reconstruct the original spatio-temporal arrangement of impacts of the original raindrops, and to be fairly confident that the pattern had actually arisen from discrete impacts of this nature.

It had seemed to me that some analysis of the CMB in this kind of statistical way ought to be able to provide a good test of the CCC proposal. So, having the occasion to visit Princeton University, at the beginning of May 2008, I took the opportunity to consult David Spergel, who is a world expert in the analysis of CMB data. I asked him if anyone had seen such an effect in the CMB data, to which he replied 'No', following this up with 'but then nobody has ever looked!' He later presented the problem to one of his post-doctoral assistants, Amir Hajian, who subsequently carried out a preliminary analysis on the observational data from the WMAP satellite observatory, to try to see if there is any evidence for this kind of effect.

What Hajian did was to choose a succession of alternative radii, starting at an angular radius of about 1° and then increasing this radius in steps of around 0.4° up to an angular radius of about 60° (for 171 different radii in all) For each given radius, circles of this radius centred on 196608 different points scattered uniformly over the sky would each have the average CMB temperature around the circle calculated. Then a histogram would be produced, to see if there is any significant deviation from what would be expected from the 'Gaussian behaviour' of completely random data. At first, certain 'spikes' were seen, seeming to present clear evidence of a number of individual circles of the nature predicted by CCC. However, before long it became clear that these were completely spurious, as the circles in question passed through certain regions of the sky, some connected with the positioning of our own Milky Way Galaxy, that were known to be hotter or colder than the normal CMB sky. To eliminate such spurious effects, information from regions close to the galactic plane had to be suppressed, and by this means the spurious 'spikes' were effectively eliminated.

A point that is worth making at this stage is that, in any case, a good many of the circles that provided the spikes had radii of over 30° in the sky, and should not have arisen in any case, according to CCC—if the aeon prior to ours had a roughly similar overall history to that which is anticipated for our own aeon. The reason for this is that the galactic black-hole encounters being considered here should not have arisen before around what would have been 'the present time' in the previous aeon, which in our aeon occurs about ⅔ of the way up the conformal diagram (Fig. 3.28). Simple geometry then shows that black-hole encounters, with *e* occurring later than ⅔ of the way up the conformal diagram of the previous aeon, would necessarily give rise to circles of radii *less* than 30° from our vantage point at *u* (in disagreement with many of the spikes). Accordingly, the temperature correlations that these effects could produce would not stretch across the celestial sphere by as much as 60°. It is a curious fact that correlations in the observed CMB temperature *do* seem to fall away at around 60°, which is unexplained in the standard inflationary picture, as far as I am aware, and this may perhaps be considered to represent some support for the CCC proposal.

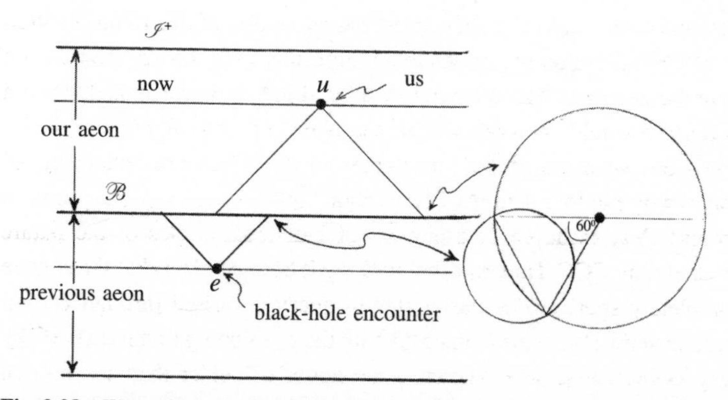

Fig. 3.28 We appear to be about ⅔ of the way up our aeon, in a conformal diagram. If this applies also to the earliest black-hole encounters in the previous aeon, then a cut-off in angular correlations at 60° is to be expected.

With the removal of these spikes, there still appeared to remain various seemingly significant systematic departures from the Gaussian randomness in Hajian's analysis. Such a departure, involving an apparent excess of cold circles in a range of angular radii between about 7° and 15°, looked particularly noteworthy and, in my opinion, required explanation. It could well be that these effects are the result of some spurious ingredients that are nothing to do with CCC, but it seemed to me that a crucial issue was whether the departures from randomness had specifically to do with the fact that the regions of the sky being averaged over were actually *circles* as opposed to some other shape, as the actual circular nature of the presumed disturbances in the CMB would appear to be a characteristic feature of this prediction of CCC. Accordingly, I suggested that the analysis be repeated, but with an area-preserving 'twist' applied to the celestial sphere (see Fig. 3.29), so that actual circles in the celestial sphere would appear to be more elliptically shaped according to the analysis. I had proposed that three different versions of the analysis should be carried out, one with no celestial twist, one with a small twist, and one with a larger twist. I had anticipated that CCC should predict that the non-Gaussian effect should be greatest with no twist, somewhat reduced with a small twist, and perhaps wiped out altogether with the large twist.

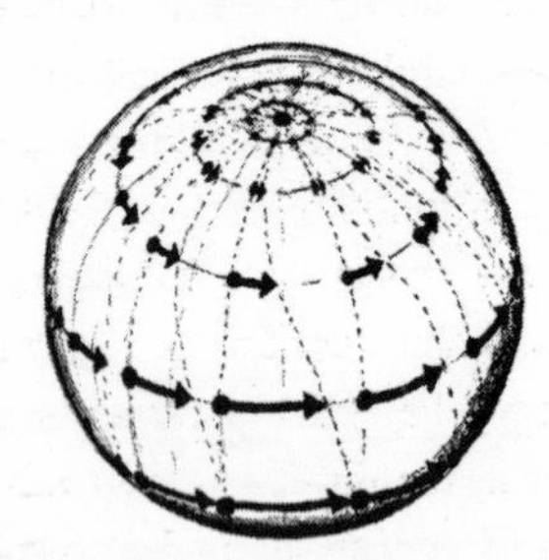

Fig. 3.29 Twisting the CMB sky (using the formula $\theta' = \theta$, $\emptyset' = \emptyset + 3a\pi\, \theta^2 - 2a\theta^3$) in spherical polar coordinates. This sends circles into more elliptical shapes.

However, the result of this analysis (carried out by Hajian in the autumn of 2008) surprised me! Completely systematically over the range of radii from 8.4° to 12.4° (which encompassed 12 successive distinct histograms), the small amount of celestial twist actually *enhanced* this particular effect very clearly, whereas the larger celestial twist did indeed make it go away. In other parts of the histograms there were somewhat similar indications of a sensitivity to the circularity of the shape being examined. At first, I was somewhat dumbfounded by this finding, being unable to imagine how the enhancement due to the small amount of twist could be explained, but then the possibility occurred to me that there might be large inhomogeneities in the mass distribution (preferably) in our own aeon that serve to distort circular images slightly into elliptical ones.[3.86] We recall, from §2.6, the significant distortions of images that the presence of Weyl curvature can produce (see Fig. 2.48). The enhancing of the effect that the small twist had produced could arise (according to my suggested picture of things) from a fortuitous agreement, in some regions of the sky, between the amount of artificial celestial twist that had been introduced and actual distortion due to Weyl curvature. In other regions the twist would lead to greater *dis*agreement, but the effect could well be an overall enhancement, in appropriate circumstances, as those due to disagreement could easily be lost in the 'noise'.

The likely presence of significant distortions, due to intervening Weyl curvature, unfortunately complicates the analysis considerably. It might be useful to break up the celestial sky into smaller regions, in order to try to identify where there might be significant Weyl curvature along the line of sight between u and the decoupling 3-surface $\mathscr{D}$. Perhaps this could be related to known inhomogeneities in the mass distribution in the universe (e.g. the large 'voids'[3.87]). In any case, there is something distinctly tantalizing about the situation in which the observations seem to have left us for the time being. It is certainly to be hoped that these matters will be clarified in the not-too-distant future, so that before too long the physical status of conformal cyclic cosmology can actually be resolved in a clear-cut way.

Epilogue

Tom looked incredulously at Aunt Priscilla, then he said "And *that*'s the craziest idea I ever heard of!"

Tom strode off to make his way to his aunt's car which would drive him home, and his aunt followed a short way behind. But presently he paused, to examine the raindrops falling on a large pond, to one side of the mill. The rain had by now tailed off considerably, to form a faint drizzle, and the impacts of individual raindrops were now clearly seen. Tom watched them for a while—and he couldn't help himself wondering …

Appendix A: Conformal rescaling, 2-spinors, Maxwell and Einstein theory

Most of the detailed equations that I give here take advantage of the 2-spinor formalism. This is not a matter of necessity, since the more familiar 4-tensor description could have been provided pretty well throughout, as an alternative. However, not only is the 2-spinor formalism simpler when it comes to expressing conformal invariance properties (see A6), but it also provides a more systematic overview when it comes to understanding the propagation of massless fields and the corresponding Schrödinger equation for their constituent particles.

Conventions employed here, including the use of abstract indices, are as in Penrose and Rindler (1984, 1986),[A.1] except that Λ denotes the cosmological constant here, rather than the 'λ' of that work, and the scalar curvature quantity 'Λ' that appears there would be $\frac{1}{24}R$. References to equations starting with 'P&R' refer to that work, and in fact all the needed equations are to be found in the 1986 Volume 2. The Einstein tensor E_{ab} used here is the *negative* of the 'Einstein tensor' $R_{ab}-\frac{1}{2}Rg_{ab}$ used there (with the same sign of Ricci tensor R_{ab} as adopted there), so that the Einstein field equations become (as in §2.6 and §3.5)

$$E_{ab}=\tfrac{1}{2}Rg_{ab}-R_{ab}=8\pi GT_{ab}+\Lambda g_{ab}.$$

A1. The 2-spinor notation: Maxwell equations

The 2-spinor formalism employs quantities with abstract spinor indices (for the complex 2-dimensional spin-space) for which I use italic capital Latin letters, either unprimed (A, B, C, ...) or primed (A', B', C', ...), these being interchanged under complex conjugation. The (complexified) tangent space at each space-time point is the tensor product of the unprimed with the primed spin-space. This enables us to adopt the abstract-index identification

$$a = AA', \quad b = BB', \quad c = CC', \ldots$$

where the italic lower-case Latin index letters a, b, c, ... refer to the space-time tangent spaces. More specifically, the tangent spaces refer to indices in *upper* position and the *co*tangent spaces to indices in the *lower* position.

The anti-symmetric Maxwell field tensor F_{ab} $(= -F_{ba})$ can be expressed in 2-spinor form in terms of a symmetric 2-index 2-spinor φ_{AB} $(= \varphi_{BA})$ by

$$F_{ab} = \varphi_{AB}\ \varepsilon_{A'B'} + \overline{\varphi}_{A'B'}\ \varepsilon_{AB}$$

where ε_{AB} $(= -\varepsilon_{BA} = \overline{\varepsilon_{A'B'}})$ is the quantity defining the complex symplectic structure of spin-space and is related to the metric by the abstract-index equation

$$g_{ab} = \varepsilon_{AB}\ \varepsilon_{A'B'},$$

spinor indices being raised or lowered according to the following prescriptions (where index-ordering on the epsilons is important!)

$$\xi^A = \varepsilon^{AB}\,\xi_B, \quad \xi_B = \xi^A\,\varepsilon_{AB}, \ \eta^{A'} = \varepsilon^{A'B'}\eta_{B'}, \quad \eta_{B'} = \eta^{A'}\,\varepsilon_{A'B'}.$$

The Maxwell field equations (denoted collectively by $\nabla\mathbf{F} = 4\pi\mathbf{J}$ in §3.2), with source as the charge-current vector J^a, are

$$\nabla_{[a}\,F_{bc]} = 0, \quad \nabla_a F^{ab} = 4\pi\ J^b,$$

(where square brackets around indices denote anti-symmetrization; round brackets, symmetrization), the charge-current conservation equation being

$$\nabla_a J^a = 0.$$

These take the respective 2-spinor forms (P&R 5.1.52, P&R 5.1.54)

$$\nabla^{A'B}\varphi^A{}_B = 2\pi J^{AA'} \quad \text{and} \quad \nabla_{AA'}J^{AA'} = 0.$$

When there are no sources ($J^a = 0$), we get the free Maxwell equations (denoted by $\mathbf{\nabla F} = 0$ in §3.2)

$$\nabla^{AA'}\varphi_{AB} = 0.$$

A2. Massless free-field ('Schrödinger') equation

This last equation is the case $n = 2$ of the massless free-field equation (P&R 4.12.42), or 'Schrödinger equation'[A.2] for a massless particle of spin $\frac{1}{2}n$ (>0):

$$\nabla^{AA'}\phi_{ABC\ldots E} = 0,$$

where $\phi_{ABC\ldots E}$ has n indices and is totally symmetric

$$\phi_{ABC\ldots E} = \phi_{(ABC\ldots E)}.$$

For the case $n = 0$, the field equation is usually taken to be $\Box\phi = 0$, where the D'Alembertian operator $\Box$ is defined by

$$\Box = \nabla_a \nabla^a,$$

but in curved space-time, we need the operator ∇_a to refer to *covariant* differentiation, and the form of equation (P&R 6.8.30)

$$(\Box + \tfrac{R}{6})\phi = 0$$

will be preferred here, as it is conformally invariant, in the sense that we shall come to shortly (A6), $R = R_a{}^a$ being the scalar curvature.

A3. Space-time curvature quantities

The (Riemann–Christoffel) *curvature* tensor R_{abcd} has the symmetries

$$R_{abcd} = R_{[ab][cd]} = R_{cdab}, \; R_{[abc]d} = 0,$$

and relates to commutators of derivatives via (P&R 4.2.31)

$$(\nabla_a \nabla_b - \nabla_b \nabla_a) V^d = R_{abc}{}^d V^c.$$

This fixes the choice of sign convention for R_{abcd}. We here define the Ricci and Einstein tensors and the Ricci scalar, respectively, by

$$R_{ac} = R_{abc}{}^b, \quad E_{ab} = \tfrac{1}{2} R g_{ab} - R_{ab}, \quad \text{where} \quad R = R_a{}^a,$$

and the Weyl conformal tensor C_{abcd} is defined by (P&R 4.8.2)

$$C_{ab}{}^{cd} = R_{ab}{}^{cd} - 2\, R_{[a}{}^{[c} g_{b]}{}^{d]} + \tfrac{1}{3} R g_{[a}{}^c g_{b]}{}^d,$$

this having the same symmetries as R_{abcd} but, in addition, all traces vanish

$$C_{abc}{}^b = 0.$$

In spinor terms, we find that we can write (P&R 4.6.41)

$$C_{abcd} = \Psi_{ABCD} \varepsilon_{A'B'} \varepsilon_{C'D'} + \overline{\Psi}_{A'B'C'D'} \varepsilon_{AB} \varepsilon_{CD}$$

where the *conformal spinor* Ψ_{ABCD} is totally symmetric

$$\Psi_{ABCD} = \Psi_{(ABCD)}.$$

The remaining information in R_{abcd} is contained in the scalar curvature R and the trace-free part of the Ricci (or Einstein) tensor, the latter being encoded in the spinor quantity $\Phi_{ABC'D'}$ with symmetries and Hermiticity

$$\Phi_{ABC'D'} = \Phi_{(AB)(C'D')} = \overline{\Phi_{CDA'B'}}$$

where (P&R 4.6.21)

$$\Phi_{ABA'B'} = -\tfrac{1}{2} R_{ab} + \tfrac{1}{8} R g_{ab} = \tfrac{1}{2} E_{ab} - \tfrac{1}{8} R g_{ab}.$$

A4. Massless gravitational sources

In Appendix B, we shall be particularly concerned with the Einstein field equations when the (symmetric) source tensor T_{ab} is *trace-free*

$$T_a{}^a=0,$$

since this is appropriate for *massless* (i.e. zero rest-mass) sources, telling us that the spinor-indexed quantity $T_{ABA'B'}=\overline{T}_{A'B'AB}=T_{ab}$ has the symmetry

$$T_{ABA'B'}=T_{(AB)(A'B')}.$$

The divergence equation $\nabla^a T_{ab}=0$, i.e. $\nabla^{AA'}T_{ABA'B'}=0$, can be re-expressed

$$\nabla_{B'}^{A'}T_{CDA'B'}=\nabla_{(B}^{A'}T_{CD)A'B'}.$$

The Einstein equations above are now (P&R 4.6.32)

$$\Phi_{ABA'B'}=4\pi G T_{ab},\ R=4\Lambda.$$

When rest-mass *is* present, so that T_{ab} has a trace

$$T_a{}^a=\mu,$$

then Einstein's equations take the form

$$\Phi_{ABA'B'}=4\pi G T_{(AB)(A'B')},\quad R=4\Lambda+8\pi G\mu.$$

A5. Bianchi identities

The general Bianchi identity $\nabla_{[a}R_{bc]de}=0$, in spinor-indexed form, becomes (P&R 4.10.7, 4.10.8)

$$\nabla_{B'}^{A}\Psi_{ABCD}=\nabla_{(B}^{A'}\Phi_{CD)A'B'}\quad\text{and}\quad\nabla^{CA'}\Phi_{CDA'B'}+\tfrac{1}{8}\nabla_{DB'}R=0.$$

When R is a constant—a situation that arises with Einstein's equations when the sources are *massless*—we have

$$\nabla^{CA'}\Phi_{CDA'B'}=0,\quad\text{whence}\quad\nabla_{B'}^{A}\Psi_{ABCD}=\nabla_{B}^{A'}\Phi_{CDA'B'},$$

the symmetry in BCD on the right being implied. Incorporating the Einstein equation, with massless sources, we get

$$\nabla^{A}_{B'}\Psi_{ABCD}=4\pi G\nabla^{A'}_{B}T_{CDA'B'}$$

(see P&R 4.10.12). Note that when $T_{ABC'D'}=0$, we obtain the equation (P&R 4.10.9)

$$\nabla^{AA'}\Psi_{ABCD}=0,$$

which is the massless free-field equation in A2, for the case $n=4$ (i.e. for spin 2).

A6. Conformal rescalings

In accordance with the conformal rescaling (with $\Omega>0$ smoothly varying)

$$g_{ab}\mapsto\hat{g}_{ab}=\Omega^2\,g_{ab},$$

we adopt the abstract-index relations

$$\hat{g}^{ab}=\Omega^{-2}\,g^{ab},$$
$$\hat{\varepsilon}_{AB}=\Omega\,\varepsilon_{AB},\quad \hat{\varepsilon}^{AB}=\Omega^{-1}\varepsilon^{AB}$$
$$\hat{\varepsilon}_{A'B'}=\Omega\,\varepsilon_{A'B'},\quad \hat{\varepsilon}^{A'B'}=\Omega^{-1}\varepsilon^{A'B'}.$$

The operator ∇_a now must transform

$$\nabla_a\mapsto\hat{\nabla}_a$$

so that the action of ∇_a on a general quantity written with spinor indices is generated by

$$\hat{\nabla}_{AA'}\phi=\nabla_{AA'}\phi,\quad \hat{\nabla}_{AA'}\xi_B=\nabla_{AA'}\xi_B-\Upsilon_{BA'}\xi_A,\quad \hat{\nabla}_{AA'}\eta_{B'}=\nabla_{AA'}\eta_{B'}-\Upsilon_{AB'}\eta_{A'},$$

where

$$\Upsilon_{AA'}=\Omega^{-1}\,\nabla_{AA'}\Omega=\nabla_a\log\,\Omega,$$

the treatment of a quantity with many lower indices being built up from these rules, one term for each index. (Upper indices have a corresponding treatment, but this will not be needed here.)

We choose the scaling for a massless field $\phi_{ABC...E}$ to be

$$\hat{\phi}_{ABC...E} = \Omega^{-1}\phi_{ABC...E}$$

and then find, applying the above prescriptions, that

$$\hat{\nabla}^{AA'}\hat{\phi}_{ABC...E} = \Omega^{-3}\nabla^{AA'}\phi_{ABC...E}$$

so that the vanishing of either side implies the vanishing of the other, whence satisfaction of the massless free-field equations is conformally invariant. In the case of the Maxwell equations with sources, we find that conformal invariance of the whole system $\nabla^{A'B}\varphi^{A}{}_{B}=2\pi J^{AA'}$, $\nabla_{AA'}J^{AA'}=0$ (P&R 5.1.52, P&R 5.1.54 in A2) is preserved with the scalings

$$\hat{\varphi}_{AB}=\Omega^{-1}\varphi_{AB} \quad \text{and} \quad \hat{J}^{AA'}= \Omega^{-4}J^{AA'},$$

since we find

$$\hat{\nabla}^{A'B}\hat{\varphi}^{A}{}_{B} = \Omega^{-4}\,\nabla^{A'B}\varphi^{A}{}_{B} \quad \text{and} \quad \hat{\nabla}^{AA'}\hat{J}_{AA'} = \Omega^{-4}\,\nabla^{AA'}J_{AA'}.$$

A7. Yang–Mills fields

It is important to observe that the Yang–Mills equations, that form the basis of our current understanding of both the strong and the weak forces of particle interactions, are also conformally invariant, so long as we can ignore the introduction of *mass* which may be taken to be through the subsequent agency of the Higgs field. The Yang–Mills field strengths can be described by a tensor quantity (a 'bundle curvature')

$$F_{ab\Theta}{}^{\Gamma}=-F_{ba\Theta}{}^{\Gamma},$$

where the (abstract) indices Θ, Γ, ... refer to the internal symmetry group (U(2), SU(3), or whatever) of relevance to the particle symmetries. We can represent this bundle curvature in terms of the spinor quantity $\varphi_{AB\Theta}{}^{\Gamma}$ (P&R 5.5.36) by

$$F_{ab\Theta}{}^{\Gamma}=\varphi_{AB\Theta}{}^{\Gamma}\varepsilon_{A'B'}+\overline{\varphi}_{A'B'}{}^{\Gamma}{}_{\Theta}\varepsilon_{AB}$$

where, for a unitary internal group, the complex conjugate of a lower internal index becomes an upper internal index, and vice versa. The field equations mirror those for the Maxwell equations, where we supply the additional internal indices as indicated above. Accordingly the conformal invariance of Maxwell theory also applies to the Yang–Mills equations, since the internal indices Θ, Γ, ... are unaffected by the conformal rescaling.

A8. Scaling of zero rest-mass energy tensors

It should be noted that for an energy tensor T_{ab} that is trace-free ($T_a{}^a=0$), we find that the scaling (P&R 5.9.2)

$$\hat{T}_{ab}=\Omega^{-2}\,T_{ab}$$

preserves the conservation equation $\nabla^a T_{ab}=0$, since we find

$$\hat{\nabla}^a\hat{T}_{ab}=\Omega^{-4}\nabla^a T_{ab}.$$

In Maxwell theory, we have an expression for the energy tensor in terms of F_{ab} that translates into spinor form as (P&R 5.2.4)

$$T_{ab}=\frac{1}{2\pi}\,\varphi_{AB}\overline{\varphi}_{A'B'}.$$

In the case of Yang–Mills theory, we simply have extra indices

$$T_{ab}=\frac{1}{2\pi}\,\varphi_{AB\Theta}{}^{\Gamma}\,\overline{\varphi}_{A'B'}\Phi_{\Gamma}.$$

For a massless scalar field, subject to the equation $(\Box+\frac{R}{6})\phi=0$ considered earlier (P&R 6.8.30), we have the conformal invariance (P&R 6.8.32)

$$(\hat{\Box}+\tfrac{\hat{R}}{6})\hat{\phi}=\Omega^{-3}(\Box+\tfrac{R}{6})\phi,$$

where

$$\hat{\phi}=\Omega^{-1}\phi,$$

and then its (sometimes called 'new improved')[A.3] energy tensor (P&R 6.8.36)

$$T_{ab}=C\{2\nabla_{A(A'}\phi\nabla_{B')}\phi-\phi\nabla_{A(A'}\nabla_{B')}\phi+\phi^2\ \Phi_{ABA'B'}\}$$

$$=\tfrac{1}{2}C\{4\nabla_a\phi\nabla^a\phi-g_{ab}\nabla_c\phi\nabla^c\phi-2\phi\nabla_a\nabla_b\phi+\tfrac{1}{6}R\phi^2g_{ab}-\phi^2R_{ab}\},$$

C being a positive constant, satisfies the required conditions

$$T_a{}^a=0,\quad \nabla^aT_{ab}=0,\quad \text{and}\quad \hat{T}_{ab}=\Omega^{-2}T_{ab}.$$

A9. Weyl tensor conformal scalings

The conformal spinor Ψ_{ABCD} encodes the information of the conformal curvature of space-time, and it is conformally invariant (P&R 6.8.4)

$$\hat{\Psi}_{ABCD}=\Psi_{ABCD}.$$

We note the curious (but important) discrepancy between this conformal invariance and that needed to preserve satisfaction of the massless free field equations, where there would be a factor Ω^{-1} on the right. To accommodate this discrepancy, we can define a quantity ψ_{ABCD} which is everywhere proportional to Ψ_{ABCD}, but which scales according to

$$\hat{\psi}_{ABCD}=\Omega^{-1}\psi_{ABCD}$$

and we find that our 'Schrödinger equation' for gravitons[A.4] (P&R 4.10.9)

$$\nabla^{AA'}\psi_{ABCD}=0.$$

in vacuum ($T_{ab}=0$), is conformally invariant. In §3.2, the above equation is written

$$\nabla\mathbf{K}=0.$$

Corresponding to the Weyl tensor C_{abcd}, above (A3, P&R 4.6.41), we can define

$$K_{abcd}=\psi_{ABCD}\ \varepsilon_{A'B'}\ \varepsilon_{C'D'}+\overline{\psi}_{A'B'C'D'}\ \varepsilon_{AB}\ \varepsilon_{CD}$$

and we find the corresponding scalings (written $\hat{\mathbf{C}}=\Omega^2\mathbf{C}$ and $\hat{\mathbf{K}}=\Omega\mathbf{K}$ in §3.2)

$$\hat{C}_{abcd}=\Omega^2C_{abcd},\quad \hat{K}_{abcd}=\Omega K_{abcd}.$$

Appendix B: Equations at crossover

As in Appendix A, conventions including the use of abstract indices are as in Penrose and Rindler (1984, 1986), but with the cosmological constant here denoted by Λ, rather than by the 'λ' of that work, the scalar curvature quantity there referred to as 'Λ' being $\frac{1}{24}R$. There are some aspects of the detailed analysis presented in what follows that are incomplete and provisional, and it is likely that refinements of these proposals will be needed for a more complete treatment. Nevertheless, we do appear to have well-defined classical equations that allow us to propagate in a consistent and fully determined way from the remote future of one aeon into the post-big-bang region of the next.

B1. The metrics $\hat{g}_{ab}$, $\mathfrak{g}_{ab}$, and $\check{g}_{ab}$

We examine the geometry in the neighbourhood of a crossover 3-surface $\mathscr{X}$, in accordance with the ideas of Part 3, where it is assumed that there is a *collar* $\mathscr{C}$, of smooth conformal space-time containing $\mathscr{X}$, which extends both to the past and to the future of $\mathscr{X}$, within which only massless fields are present within $\mathscr{C}$ prior to the crossover $\mathscr{B}$. We choose a smooth metric tensor $\mathfrak{g}_{ab}$ in this collar, consistent with the given conformal structure—locally at least and in an initially somewhat arbitrary way. Let Einstein's *physical* metric in the 4-region $\mathscr{C}^\wedge$, just prior to $\mathscr{X}$ be $\hat{g}_{ab}$, and in the 4-region $\mathscr{C}^\vee$ immediately following $\mathscr{X}$ be $\check{g}_{ab}$, where

$$\hat{g}_{ab} = \Omega^2 \mathsf{g}_{ab} \quad \text{and} \quad \check{g}_{ab} = \omega^2 \mathsf{g}_{ab}.$$

(Note that these are not quite the conventions used in §3.2, since there the 'unhatted' g_{ab} was used for Einstein's physical metric. The explicit formulae given in Appendix A remain valid here as they stand, however.) As a 'mnemonic', we may relate the symbols '$\wedge$' and '$\vee$' to the respective portions of the null cones at points of $\mathscr{X}$. In each of these two regions we are to assume that Einstein's equations hold, with a fixed cosmological constant Λ, and that all gravitational sources in the *earlier* region $\mathscr{C}^\wedge$ are taken to be *massless*, so that their total energy tensor $\hat{T}_{ab}$ is *trace-free*

$$\hat{T}_a{}^a = 0.$$

For reasons that will emerge later, I shall use a different letter $\check{U}_{ab}$ for the energy tensor in $\mathscr{C}^\vee$, and it turns out, for consistency with the formalism, that this tensor actually has to acquire a small trace

$$\check{U}_a{}^a = \mu,$$

so that a rest-mass component to the energy tensor begins to emerge in $\mathscr{C}^\vee$. It may be conjectured that this has something to do with the emergence of rest-mass in accordance with the Higgs mechanism,[B.1] but this idea is not explored here. (It should be noted that the 'hatted' quantities such as $\hat{T}_{ab}$ have their indices raised and lowered respectively by $\hat{g}^{ab}$ and $\hat{g}_{ab}$ or, correspondingly, $\hat{\varepsilon}^{AB}$, $\hat{\varepsilon}^{A'B'}$, $\hat{\varepsilon}_{AB}$, and $\hat{\varepsilon}_{A'B'}$, whereas the 'reverse-hatted' quantities such as $\check{U}_{ab}$ would use $\check{g}^{ab}$, $\check{g}_{ab}$, $\check{\varepsilon}^{AB}$, $\check{\varepsilon}^{A'B'}$, $\check{\varepsilon}_{AB}$, and $\check{\varepsilon}_{A'B'}$). The Einstein equations hold in the respective regions $\mathscr{C}^\wedge$ and $\mathscr{C}^\vee$, so we have 'hatted' and 'reverse-hatted' versions holding:

$$\hat{E}_{ab} = 8\pi G\hat{T}_{ab} + \Lambda\hat{g}_{ab},$$
$$\check{E}_{ab} = 8\pi G\check{U}_{ab} + \Lambda\check{g}_{ab},$$

where I assume that the *same*[B.2] cosmological constant holds in the two regions, so that

$$\hat{R} = 4\Lambda, \quad \check{R} = 4\Lambda + 8\pi G\mu.$$

For the moment, the metric $\mathfrak{g}_{ab}$, which straddles the cross-over 3-surface $\mathscr{X}$, is chosen completely freely, but smoothly and consistently with the given conformal structures of $\mathscr{C}^\wedge$ and $\mathscr{C}^\vee$. Later, I provide a proposal which appears to fix a unique scaling for $\mathfrak{g}_{ab}$ in a canonical and appropriate way, so that ultimately a specific choice of $\mathfrak{g}_{ab}$ is provided, for which I propose the notation 'g_{ab}' in standard italic type. I also use standard italic type for the curvature quantities R_{abcd}, etc. whether or not the specialization of $\mathfrak{g}_{ab}$ to g_{ab} is made.

B2. Equations for $\mathscr{C}^\wedge$

In what follows, I first consider equations relating to the region $\mathscr{C}^\wedge$ and deal with $\mathscr{C}^\vee$ afterwards (see B11). We can express the transformation law of the Einstein (and Ricci) tensor as (P&R 6.8.24)

$$\hat{\Phi}_{ABA'B'} - \Phi_{ABA'B'} = \Omega \nabla_{A(A'} \nabla_{B')B} \Omega^{-1} = -\Omega^{-1} \hat{\nabla}_{A(A'} \hat{\nabla}_{B')B} \Omega$$

together with (P&R 6.8.25)

$$\Omega^2 \hat{R} - R = 6\ \Omega^{-1} \Box \Omega,$$

i.e.

$$(\Box + \tfrac{R}{6})\Omega = \tfrac{1}{6} \hat{R}\ \Omega^3.$$

This last equation has considerable pure-mathematical interest, being an instance of what is referred to as the *Calabi equation*.[B.3] But it also has *physical* interest, being the equation for a conformally invariant self-coupled scalar field ϖ which, with $\hat{R} = 4\Lambda$, we can write as

$$(\Box + \tfrac{R}{6})\varpi = \tfrac{2}{3}\Lambda\ \varpi^3.$$

Every solution of this 'ϖ-equation', as I shall henceforth refer to it, provides us with a new metric $\varpi^2 \mathfrak{g}_{ab}$ whose scalar curvature has the constant value 4Λ. The conformal invariance of the ϖ-equation is expressed in the fact that if we choose a *new* conformal factor $\tilde{\Omega}$ and transform from g_{ab} to a new conformally related metric $\tilde{g}_{ab}$

$$g_{ab} \mapsto \tilde{g}_{ab} = \tilde{\Omega}^2 g_{ab}$$

then the conformal scaling for the ϖ-field

$$\tilde{\varpi} = \tilde{\Omega}^{-1}\varpi$$

gives us (as has been remarked upon earlier, in A8; see P&R 6.8.32)

$$(\tilde{\Box} + \tfrac{\tilde{R}}{6})\tilde{\varpi} = \tilde{\Omega}^{-3}(\Box + \tfrac{R}{6})\varpi,$$

from which the required conformal invariance of the non-linear ϖ-equation immediately follows. (Note that when $\tilde{\Omega} = \Omega$ and $\varpi = \Omega$ we simply revert to Einstein's $\hat{g}_{ab}$ metric, with $\tilde{\varpi} = 1$, and the equation becomes the identity $\tfrac{2}{3}\Lambda = \tfrac{2}{3}\Lambda$.)

We have seen in A8 that the *energy tensor* for such a physically regarded ϖ-field, when the ϖ^3 term is *absent*, would be (P&R 6.8.36)

$$\begin{aligned} T_{ab}[\varpi] &= C\{2\nabla_{A(A'}\varpi\, \nabla_{B')B}\varpi - \varpi\nabla_{A(A'}\nabla_{B')B}\varpi + \varpi^2\, \Phi_{ABA'B'}\} \\ &= C\varpi^2\{\varpi\nabla_{A(A'}\nabla_{B')B}\varpi^{-1} + \Phi_{ABA'B'}\} \end{aligned}$$

where C is some constant. Moreover, we find that the ϖ^3 term in the ϖ-equation does not disturb the conservation equation $\nabla^a T_{ab}[\varpi] = 0$, so we adopt this expression for the energy tensor for the ϖ-field also, and for consistency with what follows, I shall choose

$$C = \tfrac{1}{4\pi G}.$$

Comparing this with (P&R 6.8.24, B2) above, we find

$$T_{ab}[\Omega] = \tfrac{1}{4\pi G}\Omega^2\, \hat{\Phi}_{ABA'B'} = \Omega^2 \hat{T}_{ab}$$

from Einstein's equation

$$\hat{\Phi}_{ABA'B'} = 4\pi G \hat{T}_{ab},$$

which holds for the $\hat{g}_{ab}$ metric. For a trace-free energy tensor, we find that the scaling $\hat{T}_{ab} = \Omega^{-2} T_{ab}$ (A8, P&R 5.9.2) preserves the conservation equation, so we are led to the somewhat remarkable re-formulation of Einstein's theory, as referred to the $\mathfrak{g}_{ab}$ metric, for massless sources T_{ab}:

$$T_{ab} = T_{ab}[\Omega].$$

B3. The role of the phantom field

I shall refer to Ω, regarded as a particular instance of the massless self-coupled conformally invariant field ϖ, as the *phantom* field.[B.4] It does not provide us with physically independent degrees of freedom; its presence (in the $\mathfrak{g}_{ab}$-metric) simply allows us the scaling freedom that we need, so that we can rescale the physical metric to obtain a smooth metric $\mathfrak{g}_{ab}$, conformal to Einstein's physical metric, which smoothly covers each of the joins from one aeon to the next. With the aid of such metrics covering the crossover 3-surfaces, we are enabled to study in detail the specific connections between aeons, in accordance with the requirements of CCC, by using explicit classical differential equations.

The role of the phantom field is just to 'keep track' of Einstein's *actual* physical metric by telling us how to scale the metric $\mathfrak{g}_{ab}$ back to the physical one (via $\hat{g}_{ab} = \Omega^2 \mathfrak{g}_{ab}$). Then we express the satisfaction of Einstein's equations in the pre-crossover space $\mathscr{C}^\wedge$, but now written in terms of the $\mathfrak{g}$-metric, simply as $T_{ab} = T_{ab}[\Omega]$; that is to say, Einstein's field equations are expressed in the demand that the total energy tensor T_{ab} of all the physical matter fields in the space-time region $\mathscr{C}^\wedge$ (assumed massless and having the correct conformal scaling) must be equal to the energy tensor of the phantom field $T_{ab}[\Omega]$. Whereas this can be regarded as simply a reformulation of Einstein's theory (using $\mathfrak{g}_{ab}$) within the open region $\mathscr{C}^\wedge$, it is actually something more subtle. It allows us to extend our equations up to, and even beyond its future boundary surface $\mathscr{I}^+$. But in order to do this effectively, we shall need to look a little more carefully at the relevant equations governing the quantities of interest, and their expected behaviours as $\mathscr{X}$ is approached. Moreover, we shall need to understand, and then eliminate, the freedom in the initially somewhat arbitrary choice of g-metric—i.e. of conformal factor Ω—that has been chosen for the 'collar' $\mathscr{C}$ that we are concerned with.

There is, indeed, some considerable freedom in Ω, as things stand. All that has been required, so far, is that Ω be such that the g_{ab}, as obtained from Einstein's physical metric $\hat{g}_{ab}$ by $g_{ab}=\Omega^{-2}\hat{g}_{ab}$ is finite, non-zero and smooth across $\mathscr{X}$. Even to demand the existence of such an Ω may seem like a strong requirement, but there are powerful results due to Helmut Friedrich[B.5] that lead us to expect, when there is a positive cosmological constant Λ, that full freedom in the massless radiation fields in a fully expanding universe model free of massive sources, is incorporated by a smooth (spacelike) $\mathscr{I}^+$. To put this another way, we can expect to find a smooth future conformal boundary $\mathscr{I}^+$, to $\mathscr{C}^\wedge$, as a more-or-less automatic consequence of the fact that the model is indefinitely expanding, all the gravitational sources being massless fields propagating according to conformally invariant equations. It should be noted that, at this stage, there is no demand that the scalar curvature R of the g-metric even be a constant, let alone that $R=4\Lambda$, so that the conformal factor Ω^{-1} taking us back to Einstein's $\hat{g}_{ab}$ would not necessarily satisfy the ϖ-equation $(\hat{\Box}+\frac{1}{6}\hat{R})\varpi=\frac{2}{3}\Lambda\varpi^3$ in the $\hat{g}$-metric.

B4. The normal N to $\mathscr{X}$

We observe that $\Omega\to\infty$, as $\mathscr{I}^+$ $(=\mathscr{X})$ is approached from below, since the role of Ω is to scale up the finite g-metric at $\mathscr{I}^+$ by an infinite amount, to become the remote future of the earlier aeon. However, we find that the quantity

$$\omega=-\Omega^{-1}$$

approaches *zero* from below, at $\mathscr{I}^+$, in a smooth way (the minus sign being needed for what follows), and it does this so that the quantity

$$\nabla^a\omega=N^a$$

is non-zero on the cross-over 3-surface $\mathscr{X}(=\mathscr{I}^+)$, and so provides us, at points of $\mathscr{X}$, with a future-pointing timelike 4-vector **N** *normal* to $\mathscr{X}$.

The idea is to try to arrange things so that this particular 'ω' continues smoothly across $\mathscr{X}$, from $\hat{\mathscr{C}}$ into the $\check{\mathscr{C}}$ region, and with non-zero derivative, so that it actually becomes the *same* (positive) quantity 'ω' as is required for $\check{\mathscr{C}}$'s Einstein metric $\check{g}_{ab}=\omega^2 g_{ab}$ (and it is for this reason that the minus sign is needed in '$\omega=-\Omega^{-1}$'). It may be remarked that the 'normalization' condition (P&R 9.6.17)

$$g_{ab}\,N^aN^b=\tfrac{1}{3}\Lambda$$

is an *automatic* general property of conformal infinity (here $\mathscr{X}$) when there are just massless sources for the gravitational field, so that

$$\left(\frac{3}{\Lambda}\right)^{\frac{1}{2}}\mathbf{N}$$

is a *unit* normal to $\mathscr{X}$, irrespective of the particular choice of conformal factor Ω.

B5. Event horizon area

As an incidental comment, we see that from this we can readily derive the fact, noted in §3.5, that the limiting area of cross-section of any *cosmological event horizon* must be $12\pi/\Lambda$. Any event horizon (taken in the earlier aeon) is the past light cone $\mathcal{C}$ of the future end point o^+, on $\mathscr{X}$, of some immortal observer in that aeon, as in §2.5 (see Fig. 2.43). Then the limiting area of cross-section of $\mathcal{C}$ as o^+ is approached from below is $4\pi r^2$, where r (in the g-metric) is the spatial radius of the cross-section. In the $\hat{g}_{ab}$ metric, this area becomes $4\pi r^2\Omega^2$, and we readily find from the above (B4) that Ωr approaches $(\frac{1}{3}\Lambda)^{-1/2}$ in the limit, as the cross-section approaches o^+, so that our required event-horizon area is indeed $4\pi\times(3/\Lambda)=12\pi/\Lambda$. (Although this argument has been presented in the context of CCC, all that is required for it is a small degree of smoothness for the spacelike conformal infinity which, as the work of Friedrich has shown,[B.6] is a very mild assumption, when $\Lambda>0$.)

B6. The reciprocal proposal

There is of course the awkwardness, in our particular situation here, that in describing the transition from $\mathscr{C}^{\wedge}$ to $\mathscr{C}^{\vee}$ we do not have a smoothly varying quantity in either Ω or ω which describes the scaling back to *both* Einstein metrics $\hat{g}_{ab}$ and $\check{g}_{ab}$ in a uniform way. But an appropriate proposal for addressing this issue does indeed appear to be to adopt the *reciprocal proposal* $\omega = -\Omega^{-1}$, referred to above, and it is then convenient to consider the 1-form $\mathbf{\Pi}$, defined by

$$\mathbf{\Pi} = \frac{d\Omega}{\Omega^2 - 1} = \frac{d\omega}{1 - \omega^2}$$

i.e.

$$\Pi_a = \frac{\nabla_a \Omega}{\Omega^2 - 1} = \frac{\nabla_a \omega}{1 - \omega^2},$$

since this 1-form is then finite and smooth across $\mathscr{X}$ so long as we adhere to the assumptions that are implicit in the above reciprocal proposal. The quantity $\mathbf{\Pi}$ encodes the information of the metric scaling of the spacetime, albeit in a (necessarily) slightly ambiguous way.[B.7] We can integrate to obtain a parameter τ, so that

$$\mathbf{\Pi} = d\tau, \quad -\coth\tau = \Omega\ (\tau < 0), \quad \tanh\tau = \omega\ (\tau \geq 0).$$

We notice that even here there is the awkward issue of a sign change, because although $\mathbf{\Pi}$ is insensitive to the replacement of Ω by Ω^{-1}, or to the replacement of ω by ω^{-1} there is a change of sign when we pass from Ω^{-1} to ω. We might take the view that the sign of the conformal factor is irrelevant, in any case, since in the rescalings of the metric $\hat{g}_{ab} = \Omega^2 g_{ab}$ and $\check{g}_{ab} = \omega^2 g_{ab}$, the conformal factors Ω and ω appear *squared*, so that adopting the positive rather than the negative value of each of these conformal factors might be regarded as purely conventional. However, as we recall from Appendix A, there are numerous quantities that scale with Ω (or ω) *un*squared, most notably there being the discrepancy between the scalings $\hat{\Psi}_{ABCD} = \Psi_{ABCD}$ and $\hat{\psi}_{ABCD} = \Omega^{-1}\psi_{ABCD}$, leading to

$$\Psi_{ABCD} = \Omega^{-1} \psi_{ABCD} \text{ i.e. } \mathbf{C} = \Omega^{-1}\mathbf{K}$$

in the space $\mathscr{C}^{\wedge}$, since there the Einstein physical metric is $\hat{g}_{ab}$, which gives us

$$\hat{\Psi}_{ABCD} = \hat{\psi}_{ABCD} \text{ i.e. } \hat{\mathbf{C}} = \hat{\mathbf{K}}$$

(this convention differing from what was adopted in §3.2 since it is now the hatted metric in which Einstein's equations hold). Thus, when considering the smooth behaviour of quantities across $\mathscr{X}$, where both Ω and ω change sign (through ∞ and 0, respectively), we must exercise care in keeping track of the physical significance of these signs.

The specific reciprocal relationship between Ω and ω that is being invoked here is, however, dependent upon a restriction in the choice of scaling for the g_{ab} metric, namely that the condition

$$R = 4\Lambda$$

holds, in conjunction with $\hat{R} = 4\Lambda = \check{R} - 8\pi G\mu$ (see B1). This scaling is easy to arrange, locally at least, by simply choosing a new (local) metric $\tilde{g}_{ab}$ for $\mathscr{C}$, to be

$$\tilde{g}_{ab} = \tilde{\Omega}^2 \mathrm{g}_{ab}$$

where $\tilde{\Omega}$ is some smooth solution of the ϖ-equation over the cross-over. This $\tilde{g}$-metric is not yet the unique g-metric that we are looking for to cover the cross-over in a canonical way, however, since there are many possible solutions $\tilde{\Omega}$ of the ϖ-equation that could be chosen. We shall come to some further requirements shortly, that our canonical metric g_{ab} might be required to satisfy. For the moment, let us simply assume that our metric g_{ab} has been chosen to have $R = 4\Lambda$ (i.e. we re-label the above $\tilde{g}_{ab}$ as our new choice of g_{ab}). Without such a restriction as $R = 4\Lambda$, this reciprocal relationship between Ω and ω could not be precise, although for the type of conformal factor ω that we expect to find with Tod's proposal[B.8] (see end of §2.6, and §3.1, §3.2), the behaviour of the conformal factor for the case of a big bang with pure radiation as the gravitational sources, as with the Tolman radiation-filled solutions[B.9] (see §3.3), indeed behaves, in the past limit as the big bang is approached, as though it were *proportional* to the reciprocal of the smooth continuation of an Ω scale factor for a previous aeon. The choice of $R=4\Lambda$ for the metric of $\mathscr{C}$, at $\mathscr{X}$, is what fixes this proportionality factor to be (–)1. This is illustrated in the fact that the somewhat remarkable relation (coming about when we act on Π_a with the divergence operator ∇^a and then apply the ϖ-equation for Ω)

$$\Omega = \frac{\nabla^a \Pi_a}{\frac{2}{3}\Lambda - 2\Pi_b \Pi^b},$$

which arises when this restriction is placed on R, the specific choice of form Π (rather than some more general form $d\Omega/(\Omega^2 - A)$, say), depends upon this restriction that the conformal factor Ω is to propagate to become (minus) its inverse $\omega = -1/\Omega$, rather than to, say, $-A/\Omega$. Note that at $\mathscr{X}$, where $\Omega = \infty$, we must have

$$\Pi_b \Pi^b = \tfrac{1}{3}\Lambda$$

and that at $\mathscr{X}$ we also have $\Pi_a = \nabla_a \omega = N_a$, the normal vector to $\mathscr{X}$, of length $\sqrt{\Lambda/3}$, as noted earlier (P&R9.6.17).

B7. Dynamics across $\mathscr{X}$

How do we expect that our dynamical equations will allow us to propagate across $\mathscr{X}$ in an unambiguous way? I am supposing that in the remote future of the earlier aeon, Einstein's equations hold, all sources being massless and propagating according to well-defined deterministic conformally invariant classical equations. We may suppose that these are Maxwell's equations, the Yang–Mills equations without mass, and things like the Dirac–Weyl equation $\nabla^{AA'}\phi_A = 0$ (the Dirac equation in the zero-mass limit), some such particles acting as sources for the gauge fields, all these taken in the limit when rest-mass is treated as having reached zero, in accordance with §3.2. The coupling of these to the gravitational field is expressed in the equation $T_{ab} = T_{ab}[\Omega]$, where Ω is the phantom field. We know that $T_{ab}[\Omega]$ should be finite on $\mathscr{X}$, despite Ω becoming infinite there, because T_{ab} ought itself to be finite at $\mathscr{X}$, the propagation of the fields involved in T_{ab} being conformally invariant and therefore not particularly concerned with the location of $\mathscr{X}$ within $\mathscr{C}$. The proposal of CCC is that, until the situation becomes more complicated, perhaps through ordinary gravitational sources beginning to acquire rest-mass, etc. via the Higgs mechanism, or whatever alternative proposal might perhaps eventually turn out to be more accurate, these *same* conformally invariant equations for the matter sources must continue into the post-big-bang region $\mathscr{C}^{\vee}$. We shall see, however, that even with the spare situation envisaged here, we are not able to escape the appearance of rest-mass in some form, soon after $\mathscr{X}$ has been crossed (see B11).

B8. Conformally invariant D_{ab} operator

To help us to understand the physical implications for $\mathscr{C}^{\vee}$, and to see how the Einstein equations for that region will operate, let us first examine $T_{ab}[\Omega]$ explicitly:

$$T_{ab}[\Omega] = \frac{1}{4\pi G}\,\Omega^2\{\Omega\nabla_{A(A'}\nabla_{B')B}\Omega^{-1} + \Phi_{ABA'B'}\}$$

which, with $\omega = -\Omega^{-1}$, we can rewrite as

$$\{\nabla_{A(A'}\nabla_{B')B} + \Phi_{ABA'B'}\}\omega = 4\pi G\omega^3 T_{ab}[\Omega].$$

This is an interesting equation in that the 2nd-order operator

$$\mathrm{D}_{ab} = \nabla_{(A|(A'}\nabla_{B')|B)} + \Phi_{ABA'B'}$$

on the left, when acting on a scalar quantity of conformal weight 1 (where the extra symmetry over AB plays no role when the operator acts, as here, on a scalar), had been earlier pointed out to be conformally invariant by Eastwood and Rice.[B.10] In tensor terms we can write this (with the sign conventions for R_{ab} adopted here) as

$$\mathrm{D}_{ab} = \nabla_a\nabla_b - \tfrac{1}{4}g_{ab}\Box - \tfrac{1}{2}R_{ab} + \tfrac{1}{8}Rg_{ab}.$$

The quantity ω does indeed have conformal weight 1, since if g_{ab} is *further* rescaled according to

$$g_{ab} \mapsto \tilde{g}_{ab} = \tilde{\Omega}^2 g_{ab}$$

then, taking the definition of $\tilde{\omega}$, for the $\tilde{g}$-metric, to mirror that of ω, for the g-metric,

$$\tilde{g}_{ab} = \tilde{\omega}^2 \hat{g}_{ab} \quad \text{to mirror} \quad g_{ab} = \omega^2 \hat{g}_{ab},$$

we find

$$\omega \mapsto \tilde{\omega} = \tilde{\Omega}\,\omega$$

(i.e. ω has conformal weight 1). Thus,

$$\tilde{D}_{ab}\tilde{\omega} = \tilde{\Omega}\, D_{ab}\omega.$$

We can write this conformal invariance in the operator form

$$\tilde{D}_{ab} \circ \tilde{\Omega} = \tilde{\Omega} \circ D_{ab}.$$

Einstein's equations for the $\hat{g}$-metric, written in the g-metric in the terms given above

$$D_{ab}\omega = 4\pi G \omega^3 T_{ab},$$

tell us that the quantity $D_{ab}\omega$ must vanish to *third order* across $\mathscr{X}$ itself, when (as would be expected) T_{ab} is smooth across $\mathscr{X}$. In particular, the fact that $D_{ab}\omega = 0$ on $\mathscr{X}$ tells us that

$$\nabla_{A|(A'}\nabla_{B')|B}\omega \; (= -\omega\Phi_{ABA'B'}) = 0 \text{ on } \mathscr{X},$$

and we can rewrite this as

$$\nabla_{(a}N_{b)} = \tfrac{1}{4} g_{ab}\nabla_c N^c \text{ on } \mathscr{X}$$

(with $N_c = \nabla_c\omega$, as in B4 above), which tells us that the normals to $\mathscr{X}$ are 'shear-free' at $\mathscr{X}$, which is the condition for $\mathscr{X}$ to be 'umbilic' at every one of its points.[B.11]

B9. Keeping the gravitational constant positive

We can gain more insights into the interpretation of the physics that is implied for us by CCC, if we examine the interaction between the massless gravitational source fields, as described by T_{ab}, and the gravitational field (or 'graviton field') ψ_{ABCD}, as implied by the equation (P&R 4.10.12) of A5, taken in 'hatted form', and rewritten in terms of $\omega = -\Omega^{-1}$. We have

$$\nabla_{B'}^{A}(-\omega\ \psi_{ABCD}) = 4\pi G \nabla_{B}^{A'}((-\omega)^2 T_{CDA'B'}),$$

from which we derive the equivalent equation, in terms of the 'unhatted' quantities,

$$\nabla_{B'}^{A}\psi_{ABCD} = -4\pi G\{\omega\nabla_{B}^{A'}T_{CDA'B'} + 3N_{B}^{A'}T_{CDA'B'}\}.$$

We note that this equation remains well behaved as ω increases smoothly through zero (from negative to positive). This illustrates the fact that the family of partial differential equations governing the evolution of the entire system, in terms of the g-metric, does not encounter difficulties when passing through $\mathscr{X}$, from $\mathscr{C}^{\wedge}$ to $\mathscr{C}^{\vee}$.

Let us imagine that we revert to use of the original $\hat{g}$-metric when we proceed into $\mathscr{C}^{\vee}$. Then (apart from the initial 'glitch' at $\mathscr{X}$), the picture that our classical equations would provide us with, for the evolution of the space-time $\mathscr{C}^{\vee}$, would be a collapsing universe model, contracting in a reverse-exponential way, inwards from infinity, seeming to be very much like a time-reverse of what is envisaged for the remote future of our own universe. However, there is an important issue of interpretation here, because when ω changes sign, from negative to positive, the 'effective gravitational constant' (as seen particularly in the $-G\omega$ in the above formula when this first term on the right begins to dominate as ω gets larger), has *changed sign* after $\mathscr{X}$ is crossed.[B.12] The alternative interpretation that CCC presents us with is that because of considerations of physical consistency with quantum field theory, etc., this particular interpretation (with a negative gravitational constant) of the physics in the early $\mathscr{C}^{\vee}$ region cannot be properly maintained in a physical way when gravitational interactions become important. The point of view of CCC is that, instead, it becomes more appropriate, as we continue to proceed into the $\mathscr{C}^\vee$ region, to adopt the physical interpretation provided by the $\check{g}$-metric, where the now-positive conformal factor ω replaces the now-negative Ω, and the effective gravitational constant now becomes positive again.

B10. To eliminate spurious g-metric freedom

An issue that presents itself at this stage is that, according to the requirements of CCC, we want a *unique* propagation into $\mathscr{C}^\vee$. This would not be problematic were it not for the unwanted additional freedom arising from an arbitrariness in the conformal factor. As things stand, this freedom provides us with some spurious degrees of freedom, which would inappropriately influence the non-conformally-invariant gravitational dynamics of $\mathscr{C}^\vee$. These spurious degrees of freedom need to be eliminated in order that the propagation through $\mathscr{X}$ be not dependent on this additional data, undetermined by the physics of $\mathscr{C}^\wedge$. This spurious 'gauge freedom' in the choice of $\check{g}$-metric can be represented as a conformal factor $\tilde{\Omega}$ that can be applied to g_{ab} to provide us with a new metric g_{ab} (in accordance with what we had earlier):

$$g_{ab} \mapsto \tilde{g}_{ab} = \tilde{\Omega}^2 g_{ab},$$

and where, as before, we adopt

$$\omega \mapsto \tilde{\omega} = \tilde{\Omega}\omega.$$

All that we have demanded of $\tilde{\Omega}$ so far is that it be a positive-valued smoothly varying scalar field on $\mathscr{C}$ (at least in local patches), which satisfies the ϖ-equation in the g-metric—this being required in order that the scalar curvature $\tilde{R}$ remain equal to 4Λ. The ϖ-equation is a second-order hyperbolic equation of standard type, so we would expect to get a unique solution for $\tilde{\Omega}$ (for a narrow enough collar of $\mathscr{X}$) if the *value* of $\tilde{\Omega}$ and the value of its *normal derivative* were both to be specified as smooth functions on $\mathscr{X}$. This would be straightforward if we knew what these values should be chosen to be, in order to achieve some distinctive characterization for the $\tilde{g}$-metric. So the question arises: what condition on this metric can we demand, in order to eliminate these spurious degrees of freedom?

The kind of thing that we *cannot* achieve, however, would be some condition on the $\tilde{g}$-metric (perhaps together with the $\tilde{\omega}$-field) that is conformally invariant within the class of rescalings that preserve $\tilde{R}=4\Lambda$. Thus, for a trivial example, we cannot use, as one of our requirements, the demand that the scalar curvature $\tilde{R}$ of the $\tilde{g}$-metric take any value *other* than 4Λ, and the demand that it actually *does* take the value 4Λ represents no additional condition whatever on the field, and so cannot be used as a further restriction to reduce the spurious freedom that we wish to eliminate. The same would apply, a little more subtly, to a proposed demand that the squared length $\tilde{g}_{ab}\tilde{N}^a\tilde{N}^b$ of the normal vector $\tilde{N}^a=\nabla^a\tilde{\omega}$ to $\mathscr{X}$ have some particular value (indices raised and lowered using the $\tilde{g}$-metric). For if that value were chosen to be anything different from $\Lambda/3$, then (as we saw earlier; P&R 9.6.17) the condition could not be satisfied; whereas if the value is chosen actually to be $\Lambda/3$, then the condition represents no restriction at all on our spurious freedom.

Similar problems would arise also with a demand such as

$$\tilde{D}_{ab}\tilde{\omega}=0,$$

which does not represent any condition on the choice of conformal factor because of the conformal invariance property (noted earlier)

$$\tilde{D}_{ab}\omega=\tilde{\Omega}D_{ab}\omega,$$

so that $\tilde{D}_{ab}\tilde{\omega}=0$ is equivalent to $D_{ab}\omega=0$. A condition like $\tilde{D}_{ab}\tilde{\omega}=0$ would in any case not do as it stands, because there are several components, and what we require is something that reperesents just *two* conditions per point of $\mathscr{X}$ (like the specification of $\tilde{\Omega}$ and its normal derivative at each point of $\mathscr{X}$). It may be noted, moreover, that (as we have seen above) $D_{ab}\omega$ *necessarily* vanishes at $\mathscr{X}$ to 3rd order, i.e.

$$\tilde{D}_{ab}\tilde{\omega}=O(\omega^3)$$

because of the relation $D_{ab}\omega=4\pi G\omega^3T_{ab}$. However, a reasonable-looking condition that *could* be demanded would be $\tilde{N}^a\tilde{N}^b\tilde{\Phi}_{ab}=0$ on $\mathscr{X}$. More specifically, we could write this suggestion as

$$\tilde{N}^a\tilde{N}^b\tilde{\Phi}_{ab}=O(\omega).$$

We could, in fact, demand that this quantity vanish to *2nd order* on $\mathscr{X}$, i.e.

$$\tilde{N}^a\tilde{N}^b\tilde{\Phi}_{ab}=O(\omega^2)$$

which might provide us with a suitable candidate for the required *two* conditions per point of $\mathscr{X}$ that would be needed in order to fix $\tilde{\Omega}$, and hence the *g-metric* via $g_{ab}=\tilde{\Omega}^2\mathrm{g}_{ab}$. From the definition of D_{ab}, these alternative conditions would be equivalent to demanding that respectively

$$\tilde{N}^{AA'}\tilde{N}^{BB'}\tilde{\nabla}_{A(A'}\tilde{\nabla}_{B')B}\tilde{\omega}=O(\omega^2),\text{ or }O(\omega^3).$$

In tensor notation, the above two different expressions are

$$\tilde{N}^a\tilde{N}^b(\tfrac{1}{8}\tilde{g}_{ab}-\tfrac{1}{2}\tilde{R}_{ab})\quad\text{and}\quad\tilde{N}^a\tilde{N}^b(\tilde{\nabla}_a\tilde{\nabla}_b-\tfrac{1}{4}\tilde{g}_{ab}\,\tilde{\Box})\tilde{\omega},$$

where we note (dropping the tildes for the moment) that

$$\nabla_{A(A'}\nabla_{B')B}=\nabla_a\nabla_b-\tfrac{1}{4}g_{ab}\Box.$$

We also note that

$$\begin{aligned}N^{AA'}N^{BB'}\nabla_{A(A'}\nabla_{B')B}\omega&=N^aN^b\nabla_a\nabla_b\;\omega-\tfrac{1}{4}N_aN^a\Box\;\omega\\&=N^aN^b\;\nabla_aN_b-\tfrac{1}{2}N_aN^a\{\omega^{-1}(N^bN_b-\tfrac{1}{3}\Lambda)+\tfrac{1}{3}\Lambda\;\omega\}.\end{aligned}$$

This rather suggests that a reasonable alternative condition, or pair of conditions, to impose might be, respectively,

$$N^aN^b\nabla_a\,N_b=O(\omega),\quad\text{or}\quad O(\omega^2),$$

as this very much simplifies the above condition (where we note that then $N^bN_b-\frac{1}{3}\Lambda$ vanishes to 2nd or 3rd order, respectively). Conversely, *if* $N^bN_b-\frac{1}{3}\Lambda$ vanishes to 2nd order then, on $\mathscr{X}$, then

$$N^aN^b\nabla_aN_b=\tfrac{1}{2}N^a\nabla_a(N^bN_b)=\tfrac{1}{2}N^a\nabla_a(N^bN_b-\tfrac{1}{3}\Lambda)=0\quad\text{on }\mathscr{X},$$

so either of these equivalent conditions (in the form $\tilde{N}^a\tilde{N}^b\tilde{\nabla}_a\tilde{N}_b=O(\omega)$ or $\tilde{N}^b\tilde{N}_b-\frac{1}{3}\Lambda=O(\omega^2)$) can be considered alternatively as one of the required restrictions on $\tilde{\Omega}$. Note that the above expression $\Omega = \nabla^a \Pi_a / (\frac{2}{3}\Lambda - 2\Pi_b \Pi^b)$, given in B6, requires a simple pole for Ω on $\mathscr{X}$, so if the denominator vanishes to 2nd order, the numerator $\nabla^a \Pi_a$ must vanish to 1st order; indeed $\tilde{\nabla}^a \tilde{\Pi}_a = O(\omega)$ is also a reasonable form of a single condition to be imposed, and we recall from B8 that $\nabla_{(a} N_{b)} = \frac{1}{4} g_{ab} \nabla_c N^c$ on $\mathscr{X}$ whence $4N^a N^b \nabla_a N_b \text{-} N_a N^a \nabla_c N^c = O(\omega)$.

We shall be seeing in B11 below that the energy tensor U_{ab} of $\mathscr{C}^\vee$ necessarily acquires a *trace* μ, according to the procedures being adopted, this indicating the emergence of gravitational sources with rest-mass. However, we find that this trace vanishes when $3\Pi^a \Pi_a = \Lambda$. One could take the view that the CCC philosophy is best served if the presence of this rest-mass is put off for as long as possible following the big bang. Accordingly, we could well consider that demanding

$$3\tilde{\Pi}^a \tilde{\Pi}_a - \Lambda = O(\omega^3)$$

provides the appropriate two numbers per point of $\mathscr{X}$ needed to fix the g-metric. We shall actually find

$$2\pi G\mu = \omega^{-4} (1-\omega^2)^2 (3\Pi^a \Pi_a - \Lambda),$$

which becomes infinite at $\mathscr{X}$ if the zero in $3\Pi^a \Pi_a - \Lambda$ is not at least 4th order. But this is not a problem, because μ appears only in the $\check{g}$-metric, in which $\mathscr{X}$ represents the singular big bang where other infinite curvature quantities would dominate over μ if we take the zero in $3\Pi^a \Pi_a - \Lambda$ to be 3rd order.

We see that there are several alternative possibilities for the required two conditions per point of $\mathscr{X}$, which might suffice to fix $\tilde{\Omega}$, and therefore the g-metric, in a unique way. At the time of writing, I have not fully settled on what would appear to be the most appropriate (and which of these conditions are independent of which others). My preference, however, is for the third-order vanishing of $3\tilde{\Pi}^a \tilde{\Pi}_a - \Lambda$, as described above.

B11. The matter content of $\mathscr{B}^{\vee}$

To see what our equations look like physically in the post-big-bang region $\mathscr{B}^{\vee}$, we must rewrite things in terms of the 'reverse-hatted' quantities, with the metric $\check{g}_{ab}=\omega^2 g_{ab}$, with $\Omega=\omega^{-1}$. As mentioned earlier, I shall write the total post-big-bang energy tensor as U_{ab}, in order to avoid confusion with the conformally rescaled energy tensor of the (massless) matter entering $\mathscr{B}^{\vee}$ from $\mathscr{B}^{\wedge}$:

$$\check{T}_{ab}=\omega^{-2}\,T_{ab}.$$
$$=\omega^{-4}\,\hat{T}_{ab}.$$

Since $\hat{T}_{ab}$ is traceless and divergence-free, this must hold for $\check{T}_{ab}$ also (the scalings being in accordance with A8):

$$\check{T}_a{}^a=0,\quad \nabla^a\check{T}_{ab}=0.$$

We shall find that the full post-big-bang energy tensor must involve two additional divergence-free components, so that

$$\check{U}_{ab}=\check{T}_{ab}+\check{V}_{ab}+\check{W}_{ab}$$

Here, $\check{V}_{ab}$ refers to a massless field, which is to be the phantom field Ω, having now become an *actual* self-coupled conformally invariant field in the $\check{g}$-metric, since $\varpi=\Omega$ now satisfies the ϖ-equation in the $\check{g}$-metric

$$(\Box+\tfrac{R}{6})\varpi=\tfrac{2}{3}\Lambda\,\varpi^3,$$

which it must do, since the ϖ-equation is conformally invariant and is satisfied by $\varpi=-1$ in the g-metric, this becoming $\varpi=-\omega^{-1}=\Omega$ in the $\check{g}$-metric. This is reading things the opposite way around from what we did for $\mathscr{B}^{\wedge}$, where the 'phantom field' Ω was taken to be a solution of the ϖ-equation in the g-metric, and interpreted merely as the scale-factor that gets us back to the physical Einstein $\hat{g}$-metric. In *that* metric the phantom field is simply '1', and so it has no independent physical content. *Now*, we are looking at Ω as an *actual* physical field in the Einstein physical metric $\check{g}_{ab}$ and its interpretation as a conformal factor is the opposite, since it tells us how to get back to the g-metric, where in that metric the field would be '1'. For this interpretation, it is indeed essential that the conformal factors ω and Ω are reciprocals of one another—although we need also to incorporate the minus sign, so that it is really $-\Omega$ that provides the scaling from $\check{g}_{ab}$ back to g_{ab}. This reverse interpretation is consistent with the equations because it is Ω, not ω, that has to satisfy the ϖ-equation in the appropriate metric.

Accordingly, the tensor ∇_{ab} is the energy tensor of this field Ω in the $\check{g}$-metric:

$$\check{V}_{ab} = \check{T}_{ab}[\Omega].$$

We find

$$\begin{aligned} 4\pi G\check{T}_{ab}[\Omega] &= \Omega^2\{\Omega\nabla_{A(A'}\nabla_{B')B}\Omega^{-1} + \Phi_{ABA'B'}\} \\ &= \Omega^3 \mathrm{D}_{ab}\Omega^{-1} = \omega^{-3}\mathrm{D}_{ab}\omega = \omega^{-2}\mathrm{D}_{ab}1 \\ &= \omega^{-2}\,\Phi_{ABA'B'}. \end{aligned}$$

Note that the trace-free and divergence properties hold:

$$\check{V}_a{}^a = 0 \text{ and } \nabla^a\check{V}_{ab} = 0.$$

It is important to notice that the equation satisfied by ω in the g-metric is *not* the ϖ-equation, for we have seen that it is Ω, namely (-1 times) the *inverse* of ω, that satisfies this equation, whence

$$(\Box + \tfrac{R}{6})\omega^{-1} = \tfrac{2}{3}\Lambda\,\omega^{-3},$$

i.e.

$$\Box\omega = 2\omega^{-1}\,\nabla^a\omega\nabla_a\omega + \tfrac{2}{3}\Lambda\{\omega - \omega^{-1}\}.$$

Accordingly, the scalar curvature of the $\check{g}$-metric is *not* constrained to be equal to 4Λ. Instead (see B2, P&R 6.8.25, A4), we have

$$\check{R} = 4\Lambda + 8\pi G\mu,$$

with

$$\omega^2\check{R} - R = 6\omega^{-1}\Box\omega,$$

whence

$$\omega^2(4\Lambda + 8\pi G\mu) - 4\Lambda = 6\omega^{-1}\{2\omega^{-1}(\nabla^a\omega\nabla_a\omega - \tfrac{1}{3}\Lambda) + \tfrac{2}{3}\Lambda\omega\},$$

from which we deduce (see B6)

$$\begin{aligned}\mu &= \tfrac{1}{2\pi G}\omega^{-4}(1-\omega^2)^2(3\Pi^a\Pi_a - \Lambda)\\ &= \tfrac{1}{2\pi G}\{3\nabla^a\Omega\nabla_a\Omega - \Lambda(\Omega^2-1)^2\}\\ &= \tfrac{1}{2\pi G}(\Omega^2-1)^2(3\Pi^a\Pi_a - \Lambda).\end{aligned}$$

The full energy tensor $\check{U}_{ab}$ is to satisfy Einstein's equations, so we have, in addition to $\check{R}=4\Lambda+8\pi G\mu$,

$$4\pi G\check{T}_{(AB)(A'B')} = \Phi_{ABA'B'}.$$

Since neither $\check{T}_{ab}$ nor $\check{V}_{ab}$ has a trace, it falls to $\check{W}_{ab}$ to pick it up:

$$\begin{aligned}\check{U}_a{}^a &= \check{W}_a{}^a = \mu\\ &= \tfrac{1}{2\pi G}(3\Pi^a\,\Pi_a - \Lambda)(\Omega^2-1)^2\end{aligned}$$

and assuming the expressions for $\check{U}_a{}^a$, $\check{T}_{ab}$, and $\check{V}_{ab}$ given above, we can calculate $\check{W}_{ab}$ from

$$4\pi G\check{W}_{ab} = 4\pi G(\check{U}_{ab} - \check{T}_{ab} - \check{V}_{ab})$$

to obtain the following expression for $4\pi G\check{W}_{ab}$

$$\begin{aligned}&\tfrac{1}{2}(3\Pi^a\Pi_a+\Lambda)(\Omega^2-1)^2\check{g}_{ab}+(2\Omega^2+1)\Omega\nabla_{A(A'}\nabla_{B')B}\Omega\\ &-2(3\Omega^2+1)\nabla_{A(A'}\Omega\nabla_{B')B}\Omega - \Omega^4\Phi_{ab}\end{aligned}$$

which is in need of further interpretation.

B12. Gravitational radiation at $\mathscr{X}$

One feature of the infinite conformal rescaling of the metric, as we pass from $\mathscr{C}^\wedge$ (with metric $\hat{g}_{ab}$) to $\mathscr{C}^\vee$ (with metric $\check{g}_{ab}$) via $\mathscr{X}$ (with metric g_{ab}) is the way in which gravitational degrees of freedom, initially present and described in the $\hat{g}$-metric by ψ_{ABCD} (usually non-zero at $\mathscr{X}$), become transferred to other quantities in the $\check{g}$-metric. Whereas we have (A9, P&R 6.8.4)

$$\hat{\Psi}_{ABCD} = \Psi_{ABCD} = \check{\Psi}_{ABCD} = O(\omega),$$

this holding smoothly across $\mathscr{X}$, the quantity ψ_{ABCD}, describing the 'graviton field' is *discontinuous* across $\mathscr{X}$. In what follows, 'ψ_{ABCD}' is used to refer to this quanitity defined in $\mathscr{C}^{\vee}$, *prior* to crossover, or to its smooth continuation into $\mathscr{C}^{\vee}$. But the actual value of the graviton field in the region $\mathscr{C}^{\vee}$ jumps to zero at $\mathscr{X}$ (in the g_{ab} metric) because if we define $^*\psi_{ABCD}$ to describe this actual field following crossover, we have $^*\check{\psi}_{ABCD}=\check{\Psi}_{ABCD}$, whence

$$^*\psi_{ABCD}=-\omega\Psi_{ABCD}=\omega^2\psi_{ABCD}$$

in $\mathscr{C}^{\vee}$. Thus

$$^*\psi_{ABCD}=O(\omega^2),$$

so that gravitational radiation is very greatly suppressed in the big bang.

However, the degrees of freedom in the gravitational radiation described by ψ_{ABCD} in $\mathscr{C}^{\wedge}$ do make their mark on the early stages of $\mathscr{C}^{\vee}$. To see this, we note that differentiating the relation

$$\Psi_{ABCD}=-\omega\psi_{ABCD}$$

gives us

$$\nabla_{EE'}\Psi_{ABCD}=-\nabla_{EE'}(\omega\psi_{ABCD})=-N_{EE'}\psi_{ABCD}-\omega\nabla_{EE'}\psi_{ABCD},$$

so that whereas the Weyl curvature vanishes on $\mathscr{X}$, its normal derivative provides a measure of the gravitational radiation (free gravitons) out at $\mathscr{I}^{\wedge}$:

$$\Psi_{ABCD}=0,\quad N^e\nabla_e\Psi_{ABCD}=-N^eN_e\psi_{ABCD}=-\tfrac{1}{3}\Lambda\psi_{ABCD}\ \text{ on } \mathscr{X}.$$

Also, from the Bianchi identities (A5, P&R4.10.7, P&R4.10.8)

$$\nabla_{B'}^{A}\Psi_{ABCD}=\nabla_{B}^{A'}\Phi_{CDA'B'}\quad\text{and}\quad\nabla^{CA'}\Phi_{CDA'B'}=0,$$

so we have

$$\nabla_{B}^{A'}\Phi_{CDA'B'}=-N_{B}^{A}\ \psi_{ABCD}\ \text{on } \mathscr{X},$$

from which it follows that

$$N^{BB'}\nabla_{B}^{A'}\Phi_{CDA'B'}=0\ \text{on } \mathscr{X}.$$

The operator $N^{B(B'}\nabla_B^{A')}$ acts tangentially along $\mathscr{X}$ (since $N^{B(B'}N_B^{A')}=0$), so this equation represents a constraint on how $\Phi_{CDA'B'}$ behaves on $\mathscr{X}$. We also note that

$$N_{A'}^{C}\nabla_A^{D'}\Phi_{BCB'D'}=-N_{A'}^{C}N_{B'}^{D}\psi_{ABCD}$$

from which it follows that the *electric part* the normal derivative of the Weyl tensor at $\mathscr{X}$

$$N_{A'}^{C}N_{B'}^{D}\psi_{ABCD}+N_{A'}^{C'}N_{B'}^{D'}\bar{\psi}_{A'B'C'D'}$$

is basically the quantity

$$N^a\nabla_{[b}\Phi_{c]d} \text{ on } \mathscr{X}$$

while the *magnetic part*

$$\mathrm{i}N_{A'}^{C'}N_{B'}^{D'}\psi_{ABCD}-\mathrm{i}N_A^{C'}N_B^{D'}\bar{\psi}_{A'B'C'D'}$$

which is basically

$$\varepsilon^{abcd}N_a\nabla_{[b}\Phi_{c]e} \text{ on } \mathscr{X}$$

(ε^{abcd} being a skew-symmetrical Levi-Civita tensor), this being the Cotton(–York) tensor which describes the intrinsic conformal curvature of $\mathscr{X}$.[B.13]

Notes

1.1 Hamiltonian theory is a framework that encompasses all of standard classical physics and which provides the essential link to quantum mechanics. See R. Penrose (2004), *The Road to Reality*, Random House, Ch.20.

1.2 Planck's formula: $E = h\nu$. For an explanation of the symbols, see Note 2.18.

1.3 Erwin Schrödinger (1950), *Statistical thermodynamics*, Second edition, Cambridge University Press.

1.4 The term 'product' is consistent with the multiplication of ordinary integers in that the product space of an m-point space with an n-point space is an mn-point space.

1.5 In 1803 the mathematician Lazare Carnot published *Fundamental principles of equilibrium and movement* where he noted the losses of 'moment of activity', i.e. the useful work done. This was the first-ever statement of the concept of transformation of energy or entropy. Sadi Carnot went on to postulate that 'some caloric is always lost' in mechanical work. In 1854 Clausius developed the idea of 'interior work', i.e. that 'which the atoms of the body exert on each other' and 'exterior work', i.e. that 'which arises from foreign influences [to] which the body may be exposed'.

1.6 Claude E. Shannon, Warren Weaver (1949), *The mathematical theory of communication*, University of Illinois Press.

1.7 In mathematical terms, the problem comes about because macroscopic indistinguishability is not what is called *transitive*, i.e. states A and B might be indistinguishable and states B and C indistinguishable, yet with A and C distinguishable.

1.8 The 'spin' of an atomic nucleus is something which requires considerations of quantum mechanics for a proper understanding, but for a reasonable physical picture, one may indeed imagine that the nucleus is 'spinning' about some axis, as might a cricket ball or baseball. The total value of this 'spin' comes about partly from the individual spins of the constituent protons and neutrons and partly through their orbital motions about one another.

1.9 E. L. Hahn (1950), 'Spin echoes'. *Physical Review* **80**, 580–94.

1.10 J.P. Heller (1960), 'An unmixing demonstration'. *Am J Phys* **28** 348–53.

1.11 It may be, however, that in the context of black holes the entropy concept does acquire some measure of genuine objectivity. We shall be examining this issue in §§2.6 and 3.4.

2.1 Various other possible interpretations of the red shift have been put forward from time to time, one of the most popular being some version of a 'tired light' proposal, according to which the photons simply 'lose energy' as they travel towards us. Another version proposes that time progressed more slowly in the past. Such schemes turn out to be either inconsistent with other well-established observations or principles, or 'unhelpful', in the sense that they can be re-phrased as being *equivalent* to the standard expanding-universe picture, but with unusual definitions of the measures of space and time.

2.2 A. Blanchard, M. Douspis, M. Rowan-Robinson, and S. Sarkar (2003), 'An alternative to the cosmological "concordance model"'. *Astronomy & Astrophysics* **412**, 35–44. arXiv:astro-ph/0304237v2 7 Jul 2003.

2.3 This term was introduced in a BBC radio broadcast on 28 March 1949, as a somewhat derogatory description, by Fred Hoyle who had been a strong supporter of the rival 'steady state theory', see

§2.2. In this book, when referring to that particular event that apparently occurred some 1.37×10^{10} years ago, I shall adopt the capitalized form of this term 'Big Bang', but when referring to other similar occurrences which may occur either in reality or in theoretical models, I shall tend to use 'big bang' without specific capitalization.

2.4 Dark matter is not 'dark' (like the large, visibly dark *dust* regions, clearly seen from their obscuring effects), but is, more appropriately, *invisible* matter. Moreover, what is referred to as 'dark energy' is quite unlike the energy possessed by ordinary matter which, in accordance with Einstein's $E=mc^2$, has an *attractive* influence on other matter. Instead it is *repulsive*, and its effects appear, so far, to be fully in accord with the presence of something quite unlike ordinary energy, namely the *cosmological constant* introduced by Einstein in 1917, and taken into consideration by virtually all standard cosmology texts since then. This constant is indeed necessarily *constant*, and so, quite unlike energy, it has no independent degrees of freedom.

2.5 Halton Arp *and 33 others*, 'An open letter to the scientific community'. *New Scientist*, May 22, 2004.

2.6 A *pulsar* is a neutron star—an extraordinarily dense object, around 10 kilometres across, with a mass somewhat more than that of the Sun—which has an enormously strong magnetic field and rapidly rotates, sending precisely repeated bursts of electromagnetic radiation detectable here on Earth.

2.7 Curiously, Friedmann himself did not actually explicitly address the easiest case where the spatial curvature is *zero*: *Zeitschrift für Physik* **21** 326–32.

2.8 That is, apart from possible topological identifications, which do not concern us here.

2.9 In both the cases $K=0$ and $K<0$ there are topologically closed-up versions (obtained by identifying certain distant points in the spatial geometry with each other) in which the spatial geometry is finite. However, in all these situations, global spatial isotropy is lost.

2.10 A supernova is an extraordinarily violent explosion of a dying star (of mass somewhat greater than our own Sun), allowing it to achieve a brightness that, for a few days, exceeds the output of the entire galaxy within which it resides. See §2.4.

2.11 S. Perlmutter *et al.* (1999), *Astrophysical J* **517** 565. A. Reiss *et al.* (1998), *Astronomical J* **116** 1009.

2.12 Eugenio Beltrami (1868), 'Saggio di interpretazione della geometria non-euclidea', *Giornale di Mathematiche* **VI** 285–315. Eugenio Beltrami (1868), 'Teoria fondamentale degli spazii di curvatura costante', *Annali Di Mat., ser. II* **2** 232–55.

2.13 H. Bondi, T. Gold (1948), 'The steady-state theory of the expanding universe', *Monthly Notices of the Royal Astronomical Society* **108** 252–70. Fred Hoyle (1948), 'A new model for the expanding universe', *Monthly Notices of the Royal Astronomical Society* **108** 372–82.

2.14 I learnt a great deal of physics and its excitement from my close friend Dennis Sciama, a strong adherent of the steady-state model at that time, in addition to attending inspirational lectures by Bondi and Dirac.

2.15 J.R. Shakeshaft, M. Ryle, J.E. Baldwin, B. Elsmore, J.H. Thomson (1955), *Mem RAS* **67** 106–54.

2.16 Temperature measures in fundamental physics tend to be given in units of 'Kelvin' (denoted simply by the letter 'K', following the temperature measure, which refers to the number of centigrade (or Celsius) units above *absolute zero*.

2.17 Abbreviations CMBR, CBR, and MBR are also sometimes used.

2.18 For a given temperature T, Planck's formula for the black-body intensity, for frequency ν, is $2h\nu^3/(e^{h\nu/kT}-1)$, where h and k are Planck's and Boltzmann's constants, respectively.

2.19 R.C. Tolman (1934), *Relativity, thermodynamics, and cosmology*, Clarendon Press.

2.20 The local group of galaxies (the galactic cluster that includes the solar system's Milky Way galaxy) appears to be moving at about 630 km s^{-1} relative to the reference frame of the CMB. A. Kogut *et al.* (1993), *Astrophysical J* **419** 1.

2.21 H. Bondi (1952), *Cosmology*, Cambridge University Press.

2.22 A curious exception appears to be provided by volcanic vents at odd places on the ocean floor upon which colonies of strange life forms depend. Volcanic activity results from heating due to radioactive material, this having been originated in some other stars which, at some distant time in the past had spewed out such material in supernova explosions. The low-entropy role of the Sun is then taken over by such stars, but the general point made in the text remains unchanged.

2.23 Slight corrections to this equation come from, on the one hand, the small amount of heating due to radioactive material referred to in Note 2.22, and, on the other, effects coming from the burning of fossil fuels and global warming.

2.24 This general point seems to have been first made by Erwin Schrödinger in his remarkable 1944 book *What is life?*

2.25 R. Penrose (1989), *The emperor's new mind: concerning computers, minds, and the laws of physics,* Oxford University Press.

2.26 It is quite a common terminology to refer to this null cone as a 'light cone', but I prefer to reserve *that* terminology for the locus in the whole space-time that is swept out by the light rays through some event p. The *null cone*, on the other hand (in the sense being used here), is a structure defined just in the *tangent space* at the point p (i.e. *infinitesimally* at p).

2.27 To be explicit about Minkowski's geometry, we can choose some arbitrary observer rest-frame and ordinary cartesian coordinates (x,y,z) to specify the spatial location of an event, with a time coordinate t for that observer's time coordinate. Taking space and time scales so that $c=1$, we find that the null cones are given by $dt^2-dx^2-dy^2-dz^2=0$. The light cone (see Note 2.26) of the origin is then $t^2-x^2-y^2-z^2=0$.

2.28 The concept of mass being referred to here ('massive', 'massless') is that of *rest-mass* I shall return to this matter in §3.1.

2.29 As we recall from §1.3, the ordinary equations of dynamics are reversible in time, so that, as far as dynamical behaviour—as governed by the submicroscopic ingredients of a physical system—is concerned, we might equally say that causation can propagate from future to past. The notion of 'causation' used in the text is, however, in accordance with standard terminology.

2.30 Length $= \int\sqrt{g_{ij}dx^i dx^j}$ See R. Penrose (2004), *The Road to Reality*, Random House, Fig. 14.20, p. 318.

2.31 J.L. Synge (1956) *Relativity: the general theory*. North Holland Publishing.

2.32 It is the existence of this natural metric that actually undermines completely the seemingly penetrating analysis made by Poincaré, when he argued that the geometry of space is basically a conventional matter, and that Euclidean geometry, being the simplest, would therefore always be the best geometry to use for physics! See Poincaré *Science and Method* (trans Francis Maitland (1914)) Thomas Nelson.

2.33 The rest energy of a particle is its energy in the rest frame of the particle, so there is no contribution to this energy (*kinetic* energy) from the *motion* of the particle.

2.34 The 'escape velocity' is the speed, at the surface of a gravitating body, that an object needs to acquire in order that it can escape completely from that body and not fall back to its surface.

2.35 This was the quasar 3C273.

2.36 See appendix of R. Penrose (1965), 'Zero rest-mass fields including gravitation: asymptotic behaviour', *Proc. Roy. Soc.* **A284** 159–203. The argument is slightly incomplete.

2.37 The somewhat bizarre circumstance underlying this is related in my book (1989), *The emperor's new mind*, Oxford University Press.

2.38 The existence of a trapped surface is an example of what we now tend to refer to as a 'quasi-local' condition. In this case, we assert the presence of a closed spacelike topological 2-surface (normally a topological 2-sphere) whose future-pointing null normals are, at the surface, all converging into the future. In *any* space-time, there will be local patches of spacelike 2-surface whose normals have this property, so the condition is not a local one; a trapped surface occurs, however, only when such patches can join up to form a closed (i.e. of *compact* topology) surface.

2.39 R. Penrose (1965), 'Gravitational collapse and space-time singularities', *Phys. Rev. Lett.* **14** 57–9. R. Penrose (1968), 'Structure of space-time', in *Batelle Rencontres* (ed. C.M. deWitt, J.A. Wheeler), Benjamin, New York.

2.40 The only requirement that a non-singular space-time needs to have in this context—and which the 'singularity' would turn out to prevent—is what is called 'future null completeness'. This requirement is that every null geodesic can be extended into the future to an indefinitely large value of its 'affine parameter'. See S.W. Hawking, R. Penrose (1996), *The nature of space and time*, Princeton University Press.

2.41 R. Penrose (1994), 'The question of cosmic censorship', in *Black holes and relativistic stars* (ed. R.M. Wald), University of Chicago Press.

2.42 R. Narayan, J.S. Heyl (2002), 'On the lack of type I X-ray bursts in black hole X-ray binaries: evidence for the event horizon?', *Astrophysical J* **574** 139–42.

2.43 The idea of a *strict* conformal diagram was first formalized by Brandon Carter (1966) following the more relaxed descriptions of *schematic* conformal diagrams that I had been systematically using from around 1962 (see Penrose 1962, 1964, 1965). B. Carter (1966), 'Complete analytic extension of the symmetry axis of Kerr's solution of Einstein's equations', *Phys. Rev.* **141** 1242–7. R. Penrose (1962), 'The light cone at infinity', in *Proceedings of the 1962 conference on relativistic theories of gravitation, Warsaw,* Polish Academy of Sciences. R. Penrose (1964), 'Conformal approach to infinity', in *Relativity, groups and topology. The 1963 Les Houches Lectures* (ed. B.S. DeWitt, C.M. DeWitt), Gordon and Breach, New York. R. Penrose (1963), 'Asymptomatic properties of fields and space-times', *Phys. Rev. Lett.* **10** 66–8.

2.44 Coincidentally, the Polish word 'skraj' is pronounced in the same way as 'scri' and means a boundary (albeit usually of a forest).

2.45 In a time-reversed steady-state model, an astronaut, in free motion, following such an orbit, would encounter the inward motion of ambient material passing at greater and greater velocity until it reaches light speed, with infinite momentum impacts, in a finite experienced time.

2.46 J.L. Synge (1950), *Proc. Roy. Irish Acad.* **53A** 83. M.D. Kruskal (1960), 'Maximal extension of Schwarzschild metric', *Phys. Rev.* **119** 1743–5. G. Szekeres (1960), 'On the singularities of a Riemannian manifold', *Publ. Mat. Debrecen* **7** 285–301. C. Fronsdal (1959), 'Completion and embedding of the Schwarzchild solution', *Phys Rev.* **116** 778–81.

2.47 S.W. Hawking (1974), 'Black hole explosions?', *Nature* **248** 30.

2.48 The notions of a cosmological event horizon and particle horizon were first formulated by Wolfgang Rindler (1956), 'Visual horizons in world-models', *Monthly Notices of the Roy. Astronom. Soc.* **116** 662. The relation of these notions to (schematic) conformal diagrams were pointed out in R. Penrose (1967), 'Cosmological boundary conditions for zero rest-mass fields', in *The nature of time* (pp. 42–54) (ed. T. Gold), Cornell University Press.

2.49 This is meant in the sense that $\mathcal{C}^{-}(p)$ is the (future) boundary of the set of points that can be connected to an event p by a future-directed causal curve.

2.50 Following my own work showing the inevitability of singularities arising in a local gravitational collapse (see Note 2.36 for 1965 reference), referred to in §2.4, Stephen Hawking produced a series of papers showing how such results could also be obtained which apply more globally in a cosmological context (in several papers in the *Proceedings of the Royal Society* (see S.W. Hawking, G.F.R. Ellis (1973), *The large-scale structure of space-time*, Cambridge University Press). In 1970, we combined forces to provide a very comprehensive theorem covering all these types of situation: S.W. Hawking, R. Penrose (1970), 'The singularities of gravitational collapse and cosmology', *Proc. Roy. Soc. Lond.* **A314** 529–48.

2.51 I first presented this kind of argument in R. Penrose (1990), 'Difficulties with inflationary cosmology', in *Proceedings of the*

14th Texas symposium on relativistic astrophysics (ed. E. Fenves), New York Academy of Science. I have never seen a response from supporters of inflation.

2.52 D. Eardley ((1974), 'Death of white holes in the early universe', *Phys. Rev. Lett.* **33** 442–4) has argued that white holes in the early universe would be highly *unstable*. But that is not a reason for their not being part of the initial state, and it is perfectly consistent with what I am saying here. The white holes could well disappear, at various rates, just as in the opposite time direction, black holes can form, at various rates.

2.53 Compare A. Strominger, C. Vafa (1996), 'Microscopic origin of the Bekenstein-Hawking entropy', *Phys. Lett.* **B379** 99–104. A. Ashtekar, M. Bojowald, J. Lewandowski (2003), 'Mathematical structure of loop quantum cosmology', *Adv. Theor. Math. Phys.* **7** 233–68. K. Thorne (1986), *Black holes: the mebrane paradigm*, Yale University Press.

2.54 Elsewhere I have given this figure with the second exponent as '123' rather than '124', but I am now pushing the value up so as to include a contribution from the dark matter.

2.55 Dividing $10^{10^{124}}$ by $10^{10^{89}}$, we get $10^{10^{124}-10^{89}}=10^{10^{124}}$, as near as makes no difference.

2.56 R. Penrose (1998), 'The question of cosmic censorship', in *Black holes and relativistic stars* (ed. R.M. Wald), University of Chicago Press. (Reprinted *J. Astrophys.* **20** 233–48 1999)

2.57 See Appendix A3 re Ricci tensor.

2.58 Using the conventions of Appendix A.

2.59 There will, however, be non-linear effects concerning how the different effects of lenses along a line of sight 'add up'. I am ignoring these here.

2.60 A. O. Petters, H. Levine, J. Wambsganns (2001), *Singularity theory and gravitational lensing*, Birkhauser.

2.61 R. Penrose (1979), 'Singularities and time-asymmetry', in S. W. Hawking, W. Israel, *General relativity: an Einstein centenary survey*, Cambridge University Press, pp. 581–638. S. W. Goode, J. Wainwright (1985), 'Isotropic singularities in cosmological models',

Class. Quantum Grav. **2** 99–115. R. P. A. C. Newman (1993), 'On the structure of conformal singularities in classical general relativity', *Proc. R. Soc. Lond.* **A443** 473–49. K. Anguige and K. P. Tod (1999), 'Isotropic cosmological singularities I. Polytropic perfect fluid spacetimes', *Ann. Phys. N.Y.* **276** 257–93.

3.1 A. Zee (2003), *Quantum field theory in a nutshell*, Princeton University Press.

3.2 There are good theoretical reasons (to do with electric charge conservation) for believing that photons are actually strictly massless. But as far as observations are conserved, there is an upper limit of $m<3\times10^{-27}$ eV on the mass of the photon. G.V. Chibisov (1976), 'Astrofizicheskie verkhnie predely na massu pokoya fotona', *Uspekhi fizicheskikh nauk* **119** no. 3. **19** 624.

3.3 There is a common use of the term 'conformal invariance' among some particle physicists, which is much weaker than the one being used here, namely that the invariance is a mere 'scale invariance', demanded only for the far more restricted transformations $\mathbf{g}\mapsto\Omega^2\mathbf{g}$ for which Ω is a *constant*.

3.4 There can, however, be an issue with regard to what is referred to as a *conformal anomaly*, according to which a symmetry of the classical fields (here the strict conformal invariance) may not hold exactly true in the quantum context. This will not be of relevance at the extremely high energies that we are concerned with here, though it could perhaps be playing a role in the way that conformal invariance 'dies off' as rest-mass begins to be introduced.

3.5 D.J. Gross (1992), 'Gauge theory – Past, present, and future?', *Chinese J Phys.* **30** no. 7.

3.6 The Large Hadron Collider is intended to collide opposing particle beams at an energy of 7×10^{12} electronvolts (1.12 μJ) per particle, or lead nuclei at an energy of 574 TeV (92.0 μJ) per nucleus.

3.7 The issue of inflation is discussed in §§3.4 and 3.6

3.8 S. E. Rugh and H. Zinkernagel (2009), 'On the physical basis of cosmic time', *Studies in History and Philosophy of Modern Physics* **40** 1–19.

3.9 H. Friedrich (1983), 'Cauchy problems for the conformal vacuum field equations in general relativity', *Comm. Math. Phys.* **91** no. 4, 445–72. H. Friedrich (2002), 'Conformal Einstein evolution', in *The conformal structure of spacetime: geometry, analysis, numerics* (ed. J. Frauendiener, H. Friedrich) Lecture Notes in Physics, Springer. H. Friedrich (1998), 'Einstein's equation and conformal structure', in *The geometric universe: science, geometry, and the work of Roger Penrose* (eds. S.A. Huggett, L.J. Mason, K.P. Tod, S.T. Tsou, and N.M.J. Woodhouse), Oxford University Press.

3.10 An example of such an inconsistency problem is the so-called grandfather paradox in which a man travels back in time and kills his biological grandfather before the latter met the traveller's grandmother. As a result, one of the traveller's parents (and by extension the traveller himself) would never have been conceived. This would imply that he could not have travelled back in time after all, which means the grandfather would still be alive, and the traveller *would* have been conceived allowing him to travel back in time and kill his grandfather. Thus each possibility *seems* to imply its own negation, a type of logical paradox. René Barjavel (1943), *Le voyageur imprudent (The imprudent traveller).* [Actually, the book refers to *an ancestor* of the time traveller not his grandfather.]

3.11 This measure on $\mathcal{P}$ is a power of '$\mathrm{d}p \wedge \mathrm{d}x$', where $\mathrm{d}p$ refers to the momentum variable corresponding to the position variable x; see for example R. Penrose (2004), *The road to reality*, §20.2. If $\mathrm{d}x$ scales by a factor Ω, then $\mathrm{d}p$ scales by Ω^{-1}. This scale invariance on $\mathcal{P}$ holds independently of any conformal invariance of the physics being described.

3.12 R. Penrose (2008), 'Causality, quantum theory and cosmology', in *On space and time* (ed. Shahn Majid), Cambridge University Press. R. Penrose (2009), 'The basic ideas of Conformal Cyclic Cosmology', in *Death and anti-death, Volume 6: Thirty years after Kurt Gödel (1906-1978)* (ed. Charles Tandy), Ria University Press, Stanford, Palo Alto, CA.

3.13 Recent experiments at the Super-Kamiokande water Cherenkov radiation detector in Japan give a lower limit of the proton half-life of 6.6×10^{33} years.

3.14 Primarily pair annihilation; I am grateful to J.D. Bjorken for making this issue clear to me. J.D. Bjorken, S.D. Drell (1965), *Relatavistic quantum mechanics*, McGraw-Hill.

3.15 The observational situation, concerning neutrinos, at the moment, is that the *differences* between their masses cannot be zero, but the possibility of *one* of the three types of neutrino being massless seems still to be a technical possibility. Y. Fukuda *et al*. (1998), 'Measurements of the solar neutrino flux from Super-Kamiokande's first 300 days', *Phys. Rev. Lett*. **81** (6) 1158–62.

3.16 These operators are the quantities constructible from the generators of the group which *commute* with all the group elements.

3.17 H.-M. Chan and S. T. Tsou (2007), 'A model behind the standard model', *European Physical Journal* **C52**, 635-663.

3.18 Differential operators measure how the quantities that they act on vary in space-time; see the Appendices to see the explicit meanings of the '∇' operators used here.

3.19 R. Penrose (1965), 'Zero rest-mass fields including gravitation: asymptotic behaviour', *Proc. R. Soc. Lond*. **A284** 159-203.

3.20 In fact, in Appendix B1, my conventions as to whether $\mathbf{g}$ or $\hat{\mathbf{g}}$ is Einstein's physical metric will be opposite to this, so that it will be 'Ω^{-1}' that tends to zero.

3.21 This depends upon the nature of the matter at $\mathscr{B}^-$ being that of *radiation*, as in Tolman's radiation model described in §3.3, rather than the dust of Friedmann's model.

3.22 The 'differential' $d\Omega/(1-\Omega^2)$ is interpreted, according to Cartan's calculus of diffferential forms as a *1-form*, or *covector*, but its invariance under $\Omega \mapsto \Omega^{-1}$ is easily checked using standard rules of calculus: see, e.g., R. Penrose (2004), *The road to reality*, Random House.

3.23 I personally find the modern tendency to refer to 'dark energy' as contributing to the universe's matter density rather inappropriate.

3.24 Even obtaining a value that is too large by 120 orders of magnitude requires some act of faith in a 'renormalization procedure', without which the value '∞' would be obtained instead (see §3.5).

3.25 Determinations based on celestial mechanics provide constraints on the variation of G of $(\mathrm{d}G/\mathrm{d}t)/G_0 \leq 10^{-12}$/year.

3.26 R.H. Dicke (1961), 'Dirac's cosmology and Mach's principle', *Nature* **192** 440–441. B. Carter (1974), 'Large number coincidences and the anthropic principle in cosmology', in *IAU Symposium 63: Confrontation of Cosmological Theories with Observational Data,* Reidel, pp. 291–98.

3.27 A. Pais (1982), *Subtle is the Lord: the science and life of Albert Einstein*, Oxford University Press.

3.28 R.C. Tolman (1934), *Relativity, thermodynamics, and cosmology*, Clarendon Press. W. Rindler (2001) *Relativity: special, general, and cosmological*. Oxford University Press.

3.29 The strict notion of analytic continuation is described in R. Penrose (2004), *The Road to Reality,* Random House.

3.30 A so-called 'imaginary number' is a quantity *a* which squares to a negative number, such as the quantity *i*, which satisfies $i^2 = -1$. See R. Penrose (2004), *The road to reality,* Random House, §4.1.

3.31 B. Carter (1974), 'Large number coincidences and the anthropic principle in cosmology', in *IAU Symposium 63: Confrontation of Cosmological Theories with Observational Data*, Reidel, pp. 291–8. John D. Barrow, Frank J. Tipler (1988), *The anthropic cosmological principle*, Oxford University Press.

3.32 L. Susskind, 'The anthropic landscape of string theory arxiv:hep-th/0302219'. A. Linde (1986), 'Eternal chaotic inflation', *Mod. Phys. Lett.* **A1** 81.

3.33 Lee Smolin (1999), *The life of the cosmos,* Oxford University Press.

3.34 Gabriele Veneziano (2004), 'The myth of the beginning of time', *Scientific American*, May.

3.35 Paul J. Steinhardt, Neil Turok (2007), *Endless universe: beyond the big bang*, Random House, London.

3.36 See, for example, C.J. Isham (1975), *Quantum gravity: an Oxford symposium*, Oxford University Press.

3.37 Abhay Ashtekar, Martin Bojowald, 'Quantum geometry and the Schwarzschild singularity'. http://www.arxiv.org/gr-qc/0509075

3.38 See, for example, A. Einstein (1931), *Berl. Ber.* 235 and A. Einstein, N. Rosen (1935), *Phys. Rev. Ser. 2* **48** 73.

3.39 See note 2.50.

3.40 See note 3.11.

3.41 There is good evidence for some far larger black holes in some other galaxies, the present record being an absolutely enormous black hole of mass $\sim 1.8 \times 10^{10}$ M⊙, about the same mass as an entire small galaxy, but there may also be a good many galaxies whose black holes are much smaller than our $\sim 4 \times 10^{6}\ M_{\odot}$ hole. The exact figure suggested in the text is far from being crucially important for the argument. My guess would be that it is actually somewhat on the low side.

3.42 J. D. Bekenstein (1972), 'Black holes and the second law', *Nuovo Cimento Letters* **4** 737-740. J. Bekenstein (1973), 'Black holes and entropy', *Phys. Rev.* **D7**, 2333–46.

3.43 J. M. Bardeen, B. Carter, S.W. Hawking (1973), 'The four laws of black hole mechanics', *Communications in Mathematical Physics* **31** (2) 161–70.

3.44 In fact, a stationary black hole (in vacuum) needs only 10 numbers to characterize it completely: its location (3), its velocity (3), its mass (1), and its angular momentum (3), despite the vast numbers of parameters that would be needed to describe the way it was formed. Thus, these 10 macroscopic parameters would seem to label an absolutely enormous region of phase space, giving a huge entropy value by Boltzmann's formula.

3.45 http://xaonon.dyndns.org/hawking

3.46 L. Susskind (2008), *The black hole war: my battle with Stephen Hawking to make the world safe for quantum mechanics*, Little, Brown.

3.47 D. Gottesman, J. Preskill (2003), 'Comment on "The black hole final state"', hep-th/0311269. G.T. Horowitz, J. Malcadena (2003), 'The black hole final state', hep-th/0310281. L. Susskind (2003), 'Twenty years of debate with Stephen', in *The future of theoretical physics and cosmology* (ed. G.W. Gibbons *et al.*), Cambridge University Press.

3.48 It was early pointed out by Hawking that the pop itself would, technically, represent a momentary 'naked singularity' in violation of the cosmic-censorship conjecture. This is basically the reason why the hypothesis of cosmic censorship is restricted to *classical* general relativity theory. R. Penrose (1994), 'The question of cosmic censorship', in *Black holes and relativistic stars* (ed. R.M. Wald), University of Chicago Press.

3.49 James B. Hartle (1998), 'Generalized quantum theory in evaporating black hole spacetimes', in *Black Holes and Relativistic Stars* (ed. R.M. Wald), University of Chicago Press.

3.50 It is a well-known result of quantum theory referred to as the 'no-cloning theorem' which forbids the copying of an unknown quantum state. I see no reason why this should not apply here. W.K. Wootters, W.H. Zurek (1982), 'A single quantum cannot be cloned', *Nature* **299** 802–3.

3.51 S.W. Hawking (1974), 'Black hole explosions', *Nature* **248** 30. S.W. Hawking (1975), 'Particle creation by black holes', *Commun. Math. Phys.* **43**.

3.52 For Hawking's newer argument, see 'Hawking changes his mind about black holes', published online by *Nature* (doi:10.1038/news040712-12). It is based on conjectural ideas that have relations to string theory. S.W. Hawking (2005), 'Information loss in black holes', *Phys. Rev.* **D72** 084013.

3.53 Schrödinger's equation is a *complex* first-order equation, and when time is reversed, the 'imaginary' number i must be replaced by $-i$ ($i=\sqrt{-1}$); see note 3.30.

3.54 For further information, see R. Penrose (2004), *The road to reality*, Random House, Chs 21–3.

3.55 W. Heisenberg (1971), *Physics and beyond*, Harper and Row, pp. 73–6. See also A. Pais (1991), *Niels Bohr's times*, Clarendon Press, p. 299.

3.56 Dirac appears to have taken no interest in the issue of 'interpreting' quantum mechanics as it stands, in order to resolve the measurement issue, taking the view that current quantum field theory is, in any case, indeed just a 'provisional theory'.

3.57 P.A.M. Dirac (1982), *The principles of quantum mechanics*. 4th edn. Clarendon Press [1st edn 1930].

3.58 L. Diósi (1984), 'Gravitation and quantum mechanical localization of macro-objects', *Phys. Lett.* **105A** 199–202. L. Diósi (1989), 'Models for universal reduction of macroscopic quantum fluctuations', *Phys. Rev.* **A40** 1165–74. R. Penrose (1986), 'Gravity and state-vector reduction', in *Quantum concepts in space and time* (eds. R. Penrose and C.J. Isham), Oxford University Press, pp. 129–46. R. Penrose (2000), 'Wavefunction collapse as a real gravitational effect', in *Mathematical physics 2000* (eds. A. Fokas, T.W.B. Kibble, A.Grigouriou, and B.Zegarlinski), Imperial College Press, pp. 266–282. R. Penrose (2009), 'Black holes, quantum theory and cosmology' (Fourth International Workshop DICE 2008), *J. Physics Conf. Ser.* **174** 012001. doi: 10.1088/1742-6596/174/1/012001

3.59 There is always a problem, when dealing with a universe which might be spatially infinite, that *total* values of quantities like entropy would come out as infinite. This is not too important a point, however, as with the assumption of a general overall spatial homogeneity, one can work, instead, with a large 'co-moving volume' (whose boundaries follow the general flow of matter).

3.60 S.W. Hawking (1976), 'Black holes and thermodynamics', *Phys. Rev.* **D13(2)** 191. G.W. Gibbons, M.J. Perry (1978), 'Black holes and thermal Green's function', *Proc Roy. Soc. Lond.* **A358** 467–94. N.D. Birrel, P.C.W. Davies (1984), *Quantum fields in curved space,* Cambridge University Press.

3.61 Personal communication by Paul Tod.

3.62 See note 3.11.

3.63 I think that my own viewpoint with regard to the 'information loss' that gives rise to black-hole entropy differs from that frequently expressed, in that I do not regard the *horizon* as the crucial location for this (since horizons are not locally discernable, in any case), but I take the view that it is really the *singularity* that is responsible for information destruction.

3.64 See note 3.42.

3.65 W.G. Unruh (1976), 'Notes on black hole evaporation', *Phys. Rev.* **D14** 870.

3.66 G.W. Gibbons, M.J. Perry (1978), 'Black holes and thermal Green's function', *Proc Roy. Soc. Lond*, **A358** 467–94. N.D. Birrel, P.C.W. Davies (1984), *Quantum fields in curved space,* Cambridge University Press.

3.67 Wolfgang Rindler (2001), *Relativity: special, general and cosmological,* Oxford University Press.

3.68 H.-Y. Guo, C.-G. Huang, B. Zhou (2005), *Europhys. Lett.* **72** 1045–51.

3.69 It might be objected that the region covered by the Rindler observers is not the whole of $\mathbb{M}$, but this objection applies also to $\mathbb{D}$.

3.70 J. A. Wheeler, K. Ford (1995), *Geons, black holes, and quantum foam,* Norton.

3.71 A. Ashtekar, J. Lewandowski (2004), 'Background independent quantum gravity: a status report', *Class. Quant. Grav.* **21** R53–R152. doi:10.1088/0264-9381/21/15/R01, arXiv:gr-qc/0404018.

3.72 J.W. Barrett, L. Crane (1998), 'Relativistic spin networks and quantum gravity', *J. Math. Phys.* **39** 3296–302. J.C. Baez (2000), *An introduction to spin foam models of quantum gravity and BF theory.* Lect. Notes Phys. 543 25–94. F. Markopoulou, L. Smolin (1997), 'Causal evolution of spin networks', *Nucl. Phys.* **B508** 409–30.

3.73 H.S. Snyder (1947), *Phys. Rev.* **71(1)** 38–41. H.S. Snyder (1947), *Phys. Rev.* **72(1)** 68-71. A. Schild (1949), *Phys. Rev.* **73**, 414–15.

3.74 F. Dowker (2006), 'Causal sets as discrete spacetime', *Contemporary Physics* **47** 1–9. R.D. Sorkin (2003), 'Causal sets: discrete gravity', (Notes for the Valdivia Summer School), in *Proceedings of the Valdivia Summer School* (ed. A. Gomberoff and D. Marolf), arXiv:gr-qc/0309009.

3.75 R. Geroch, J.B. Hartle (1986), 'Computability and physical theories', *Foundations of Physics* **16** 533–50. R.W. Williams, T. Regge (2000), 'Discrete structures in physics', *J. Math. Phys.* **41** 3964–84.

3.76 Y. Ahmavaara (1965), *J. Math. Phys.* **6** 87. D. Finkelstein (1996), *Quantum relativity: a synthesis of the ideas of Einstein and Heisenberg*, Springer-Verlag.

3.77 A. Connes (1994), *Non-commutative geometry*, Academic Press.

S. Majid (2000), 'Quantum groups and noncommutative geometry', *J. Math. Phys.* **41** (2000) 3892–942.

3.78 B. Greene (1999), *The elegant universe*, Norton. J. Polchinski (1998), *String theory*, Cambridge University Press.

3.79 J. Barbour (2000), *The end of time: the next revolution in our understanding of the universe*, Phoenix. R. Penrose (1971), 'Angular momentum: an approach to combinatorial space-time', in *Quantum theory and beyond* (ed. T. Bastin), Cambridge University Press.

3.80 For an explanation of twistor theory, see R. Penrose (2004), *The road to reality,* Random House, ch. 33.

3.81 G. Veneziano (2004), 'The myth of the beginning of time', *Scientific American* (May). See also Note 3.34.

3.82 R. Penrose (2004), *The road to reality,* Random House, §28.4.

3.83 'Realizing' a quantum fluctuation as an actual irregularity in a classical matter distribution actually requires a manifestation of the **R**-process referred to towards the end of §3.4, which is not part of a *unitary evolution* **U**.

3.84 D.B. Guenther, L.M. Krauss, P. Demarque (1998), 'Testing the constancy of the gravitational constant using helioseismology', *Astrophys. J.* **498** 871–6.

3.85 In fact, there are standard procedures for taking into account the evolution from $\mathscr{B}^-$ to $\mathscr{D}$. This was not applied in Hajian's preliminary analysis of the CMB data (to be described shortly in the text), however.

3.86 Such distortions of circular shape could also occur in the previous aeon, though my guess is that this would be a smaller effect. In any case, if these occur, their effects would be much harder to deal with, and would be a great nuisance for the analysis, for a multitude of reasons.

3.87 V.G. Gurzadyan, C.L. Bianco, A.L. Kashin, H. Kuloghlian, G. Yegorian (2006), 'Ellipticity in cosmic microwave background as a tracer of large-scale universe', *Phys. Lett. A* **363** 121–4. V.G. Gurzadyan, A.A. Kocharyan (2009), 'Porosity criterion for hyperbolic voids and the cosmic microwave background',

Astronomy and Astrophysics **493** L61–L63 [DOI: 10.1051/000-6361:200811317]

A.1 R. Penrose, W. Rindler (1984), *Spinors and space-time, Vol. I: Two-spinor calculus and relativistic fields*, Cambridge University Press. R. Penrose, W. Rindler (1986), *Spinors and space-time, Vol. II: Spinor and twistor methods in space-time geometry*, Cambridge University Press.

A.2 P.A.M. Dirac (1982), *The principles of quantum mechanics*, 4th edn. Clarendon Press [1st edn 1930]. E.M. Corson (1953) *Introduction to tensors, spinors, and relatavistic wave equations.* Blackie and Sons Ltd.

A.3 C.G. Callan, S. Coleman, R. Jackiw (1970), *Ann. Phys. (NY)* **59** 42. E.T. Newman, R. Penrose (1968), *Proc. Roy. Soc., Ser. A* **305** 174.

A.4 This is the spin-2 Dirac–Fierz equation, in the linearized limit of general relativity. Dirac, P.A.M. (1936), Relativistic wave equations. *Proc. Roy. Soc. Lond.* **A155**, 447–59. M. Fierz, W. Pauli (1939), 'On relativistic wave equations for particles of arbitrary spin in an electromagnetic field', *Proc. Roy. Soc. Lond.* **A173** 211–32.

B.1 It may well be that the present formalism should be modified so that a decaying rest-mass in $\mathscr{C}^\wedge$, in accordance with §3.2, is also incorporated. However, this would be likely to complicate matters considerably, so for the moment I am restricting attention to situations that can be well treated with the assumption that our 'collar' contains no rest-mass in $\mathscr{C}^\wedge$.

B.2 I do not believe that $\hat{\Lambda}=\check{\Lambda}$ is, in itself, is a big assumption; it is just an issue of convenience. As things stand, it is merely a matter of arranging that any changes in physical constants that might occur from one aeon to the next are taken up by other quantities. As a further comment, it may be remarked that an alternative to the standard 'Planck units' introduced in §3.2, one might consider replacing the condition $G=1$ by $\Lambda=3$, as this fits in well with the formalism of CCC as presented here.

B.3 E. Calabi (1954), 'The space of Kähler metrics', *Proc. Internat. Congress Math. Amsterdam*, pp. 206–7.

B.4 Phantom field: this term has also been used, in the literature, in various other somewhat different senses.

B.5 See note 3.9.

B.6 See note 3.9.

B.7 The full freedom is given by the replacement $\Omega \mapsto (A\Omega+B)/(B\Omega+A)$, with A and B constant, whereby $\mathbf{\Pi} \mapsto \mathbf{\Pi}$. But this ambiguity is dealt with by the demand that Ω have a pole (and ω a zero) at X.

B.8 K.P. Tod (2003), 'Isotropic cosmological singularities: other matter models', *Class. Quant. Grav.* **20** 521–34. [DOI: 10.1088/0264-9381/20/3/309]

B.9 See note 3.28.

B.10 This operator was apparently introduced, in effect, by C.R. LeBrun ((1985), 'Ambi-twistors and Einstein's equations', *Classical Quantum Gravity* **2** 555–63) in his definition of 'Einstein bundle' in twistor theory. It forms part of a much more general family of operators introduced by Eastwood and Rice (M.G. Eastwood and J.W. Rice (1987), 'Conformally invariant differential operators on Minkowski space and their curved analogues', *Commun. Math. Phys.* **109** 207–28, Erratum, *Commun. Math. Phys.* **144** (1992) 213). It has relevance also in other contexts (M.G. Eastwood (2001), 'The Einstein bundle of a nonlinear graviton', in *Further advances in twistor theory vol III*, Chapman & Hall/CRC, pp. 36–9. T.N. Bailey, M.G. Eastwood, A.R. Gover (1994), 'Thomas's structure bundle for conformal, projective, and related structures', *Rocky Mtn. Jour. Math.* **24** 1191–217.) It has come to be known as the 'conformal to Einstein' operator. see also footnote on p.124 of R. Penrose, W. Rindler (1986), *Spinors and space-time, Vol. II: Spinor and twistor methods in space-time geometry*, Cambridge University Press.

B.11 This interpretation was pointed out to me by K.P. Tod. In Penrose and Rindler (1986), the condition is referred to as the 'asymptotic Einstein condition'. R. Penrose, W. Rindler (1986), *Spinors and space-time, Vol. II: Spinor and twistor methods in space-time geometry*, Cambridge University Press.

B.12 There are other ways of seeing this effective sign change in the gravitational constant, one being in the comparison between the 'Grgin behaviour' of the radiation field and the 'anti-Grgin behaviour' of the gravitational sources as conformal infinity is crossed; see Penrose and Rindler (1986), §9.4, pp. 329–32. R. Penrose, W. Rindler (1986), *Spinors and space-time, Vol. II: Spinor and twistor methods in space-time geometry*, Cambridge University Press.

B.13 K.P. Tod, personal communication.

Index

Note: roman numbers (ix, x) indicate material in the Preface; page numbers in *italics* refer to Figures; endnotes are numbered as suffix to the page number on which they appear, e.g. 280n[3.29]; CCC = conformal cyclic cosmology, CMB = cosmic microwave background